T0201487

THE QUANTUM INTERNET

Following the emergence of quantum computing, the subsequent quantum revolution will be that of interconnecting individual quantum computers at the global level. In the same way that classical computers only realised their full potential with the emergence of the internet, a fully realised quantum internet is the next stage of evolution for quantum computation. This cutting-edge book examines in detail how the quantum internet would evolve in practice, focusing not only on the technology itself but also on the implications it will have economically and politically, with numerous non-technical sections throughout the text providing broader context to the discussion. The book begins with a description of classical networks before introducing the key concepts behind quantum networks, such as quantum internet protocols, quantum cryptography, and cloud quantum computing. This book is written in an engaging style and is accessible to graduate students in physics, engineering, computer science and mathematics.

PETER P. ROHDE is an ARC Future Fellow and Senior Lecturer in the Centre for Quantum Software & Information at the University of Technology Sydney. His theoretical proposals have inspired several world-leading experimental efforts in optical quantum information processing.

THE QUANTUM INTERNET

The Second Quantum Revolution

PETER P. ROHDE

University of Technology Sydney

CAMBRIDGE
UNIVERSITY PRESS

CAMBRIDGE
UNIVERSITY PRESS

University Printing House, Cambridge CB2 8BS, United Kingdom

One Liberty Plaza, 20th Floor, New York, NY 10006, USA

477 Williamstown Road, Port Melbourne, VIC 3207, Australia

314–321, 3rd Floor, Plot 3, Splendor Forum, Jasola District Centre, New Delhi – 110025, India

103 Penang Road, #05–06/07, Visioncrest Commercial, Singapore 238467

Cambridge University Press is part of the University of Cambridge.

It furthers the University's mission by disseminating knowledge in the pursuit of education, learning, and research at the highest international levels of excellence.

www.cambridge.org
Information on this title: www.cambridge.org/9781108491457
DOI: 10.1017/9781108868815

First published 2021

Printed in the United Kingdom by TJ Books Limited, Padstow Cornwall

A catalogue record for this publication is available from the British Library.

ISBN 978-1-108-49145-7 Hardback

In memory of Prof Jonathan P. Dowling

Contents

Preface

Quantum technologies are not just of interest to quantum physicists but will have transformative effects across countless areas – the next technological revolution. For this reason, this work is directed at a general audience of not only preexisting quantum computer scientists but also classical computer scientists, physicists, economists, artists, musicians, and computer, software and network engineers. More broadly, we hope that this work will be of interest to those who recognise the future significance of quantum technologies and the implications (or even just curiosities) that globally networking them might have – the creation of the global quantum internet [182, 99]. We expect that the answer to that question will look very different to what emerged from the classical internet.

A basic understanding of quantum mechanics [157], quantum optics [73], quantum computing and quantum information theory [127],[1] classical networking [177] and computer algorithms [48] are helpful, but not essential, to following our discussion. Some mathematical sections require a basic understanding of the mathematical notation of quantum mechanics, although the reader without this background ought to be able to nonetheless follow the broader arguments.

The entirely technically disinterested or mathematically incompetent reader may refer to just Parts I, IX and X – essentially brief, nontechnical, highly speculative essays about the motivation, applications and implications of the future quantum internet.

This work is partially a review of existing knowledge relevant to quantum networking and partially original ideas, to a large extent based on the adaptation of classical networking concepts and quantum information theory to the context of quantum networking. A reader with an existing background in these areas could skip the respective review sections.

---

[1] Throughout this book we use the Nielsen and Chuang convention for the pronunciation of 'zed' [127].

Our goal is to present a broadly accessible technical and nontechnical overview of how we foresee quantum technologies to operate in the era of quantum globalisation and the exciting possibilities and emergent phenomena that will evolve from it.

We do not shy away from making bold predictions about the future of the quantum internet, how it will manifest itself and what its implications will be for humanity and for science. Inevitably, some of our predictions will turn out to be accurate, whereas others will completely miss the mark entirely. We have no fear of controversy. How accurate our vision will be will have to be seen, but the most important goal in presenting grandiose predictions is to inspire new research directions, encourage future work and stimulate lively and rigorous scientific debate about future technology. If we succeed at achieving these things, yet every last one of our predictions turn out to be completely and utterly wrong, we will consider this work a resounding success. Our goal, first and foremost, is to inspire future science.

Acknowledgements

The desire to share and unite remote digital assets motivated the development of the classical internet, the enabler of the entire twenty-first century economy and our modern way of life. As we enter the quantum era, it is to be expected that there will be a similar demand for networking quantum assets, motivating a *global quantum internet* for bringing together the world's quantum resources, leveraging off their exponential trajectory in capability. We present models for quantum networking, how they might be applied in the future and the implications they will have.

Like the classical internet, it is to be expected that the implications of the quantum internet will be far more than technological, with far-reaching economic, political and geostrategic consequences, which to a large extent act as the driving force for how they will evolve. Although it is impossible to make concrete predictions for the future, we present our treatment of the topic holistically, discussing the interplay between the technology and its driving forces. This includes economic and strategic game-theoretic models that are unique to these quantum technologies, with no direct analogue in terms of conventional analyses. In short, the nonlinear scaling in the utility of quantum resources requires nonlinear economic and strategic models. The nonlinear nature in the utility of future quantum infrastructure implies 'quantum enhancement' not only from a physical perspective but in terms of their implications for humanity.

This work is based on the combined efforts of a highly interdisciplinary team. Zixin Huang contributed to multiple sections of the book, in particular to those on cryptography and quantum algorithms. He-Liang Huang and Zu-En Su contributed to the sections on experimental quantum optics, the Chinese quantum satellite developments and the early structure of the book. Simon Devitt contributed the sections on the quantum SneakerNet, error correction and fault tolerance. Rohit Ramakrishnan and Chandrashekar Radhakrishnan contributed to the sections on optical interfacing and switching, quantum memories and experimental quantum optics. Si-Hui Tan and Atul Mantri contributed to the sections on secure cloud quantum computing. Nana Liu contributed on quantum machine learning and quantum algorithms. Scott Harrison contributed to the sections on economics and game theory. Tim Byrnes contributed the sections on clock synchronisation and telescopy. William J. Munro contributed the sections on quantum repeater networks and provided editorial assistance. Jonathan P. Dowling acted as co-editor, although his recent passing implies that a number of differences in editorial opinion now swing in Peter Rohde's favour, who acted as lead author and editor.

1

Introduction

The internet is one of the key technological achievements of the twentieth century, an enabling factor in every aspect of our everyday use of modern technology. Whereas digital computing was the definitive technology of the twentieth century, quantum technologies will be for the 21st [127, 23].

Perhaps the most exciting prospect in the quantum age is the development of quantum computers. Richard Feynman [65] was the first to ask the question *'If quantum systems are so exponentially complex that we are unable to simulate them on our classical computers, can those same quantum systems be exploited in a controlled way to exponentially outperform our classical computers?'* Subsequently, the Deutsch-Jozsa algorithm [52] demonstrated for the first time that algorithms can run on a quantum computer, exponentially outperforming any classical algorithm. Since then, an enormous amount of research has been dedicated to finding new quantum algorithms, and the search has indeed been a very fruitful one,[1] with many important applications having been found, including, amongst many others:

- Searching unstructured databases:
  - Grover's algorithm [83].
  - Quadratic speedup.
- Satisfiability and optimisation problems:[2]
  - Grover's algorithm.
  - Quadratic speedup.
  - Includes solving **NP**-complete problems and brute-force cracking of private encryption keys.

---

[1] See the Quantum Algorithm Zoo for a comprehensive summary of the current state of knowledge on quantum algorithms.
[2] A satisfiability problem is one in which we search a function's input space for a solution(s) satisfying a given output constraint. The hardest such problems, like the archetypal 3-SAT problem, are **NP**-complete.

- Many optimisation problems are **NP**-complete or can be approximated in **NP**-complete.

- Period finding and integer factorisation:

    - Shor's algorithm [165].
    - Exponential speedup.
    - This compromises both Rivest, Shamir and Adleman (RSA) and elliptic-curve public-key cryptography [141], the most widely used cryptographic protocols on the internet today.
    - This problem is believed to be **NP**-intermediate – an **NP** problem that lies outside **P** (and is therefore classically hard) but that is not **NP**-complete (the 'hardest' of the **NP** problems).

- Simulation of quantum systems:

    - Lloyd's algorithm [107].
    - Exponential speedup.
    - This includes simulation of molecular and atomic interactions in the study of quantum chemistry or nuclear physics; interactions between drug molecules and organic molecules for drug design; genetic interactions for the study of genetics and genetic medicine; nanoscale semiconductor physics for integrated circuit design; and much more.

- Simulation of quantum field theories:

    - Jordan-Lee-Preskill algorithm [94, 34].
    - Exponential speedup.
    - A key area of fundamental physics research.

- Topological data analysis:

    - Lloyd's algorithm [108].
    - Exponential speedup.
    - Broad applications including social media network analysis; consumer behaviour; behavioural dynamics; neuroscience; and higher-dimensional signal and image processing.

- Solving linear systems of equations:

    - Algorithms by [84, 26].
    - Exponential speedup.
    - Widespread applications in linear algebra and calculus.

- Quantum machine learning:

    - Lloyd's algorithm [109].
    - This includes putting an end to humanity.

Some of these are discussed in more detail in Chapter 28.

It is likely we have not yet begun to fully recognise the capabilities of quantum computers and the full plethora of applications they may have in the future. We stand at the beginning of the emergence of an entirely new type of technology.

In addition to many practical applications, the onset of quantum computing carries with it deep philosophical implications, specifically, the extended Church-Turing (ECT) thesis hypothesises that any physically realisable system can be *efficiently*[3] simulated by a universal Turing machine (i.e., classical computer). The believed exponential complexity of quantum systems inclines quantum computer scientists to believe that the ECT thesis is therefore false [50].[4] The demonstration of large-scale quantum computers, though unable to prove or disprove the ECT thesis,[5] could at least provide some convincing evidence against the ECT conjecture.

From a computational complexity theorist's perspective, it is strongly believed that the complexity classes of problems efficiently solvable on classical computers (**P** and **BPP**) and quantum computers (**BQP**) are distinct. Specifically, it is believed that **BPP** $\subset$ **BQP**. If this conjecture is correct, it implies the existence of quantum algorithms superpolynomially faster than the best classical ones and that the ECT thesis is not correct. More specifically, Figure 1.1 illustrates the believed relationships between some of the most important complexity classes relevant to quantum computing.

In addition to quantum computing, quantum cryptography holds the promise of uncrackable cryptographic protocols, guaranteed not by the assumed complexity of solving certain mathematical problems like integer factorisation or brute-force searching but by the laws of quantum mechanics. That is, provided that our understanding of quantum mechanics is correct, quantum cryptographic protocols exist that cannot be cracked, irrespective of the computational resources of an adversary.

Already we are beginning to see elementary realisations of essential quantum technologies such as quantum computing, cryptography and metrology. As these technologies become increasingly viable and more ubiquitous, the demand for networking them and sharing quantum resources between them will become a pressing issue. Most notable, quantum cryptography and *cloud quantum computing* will be pivotal in the proliferation of quantum technology, which necessarily requires reliable quantum communications channels.

---

[3] The term 'efficient' is one coined by the computer scientist to mean that a problem can be solved in time at most polynomial in the size of the problem.

[4] We have discovered a truly marvellous proof of this, which this footnote is too narrow to contain.

[5] When we talk about 'scalability' or the 'ECT thesis' we are talking about asymptotic relationships. Clearly no finite-sized experiment can prove asymptotic scaling with certainty. But with a sufficiently large quantum computer at our disposal, demonstrating exponentially more computational power than its classical sibling, we might be reasonably satisfied in convincing ourselves about the nature of the scaling of different computational models.

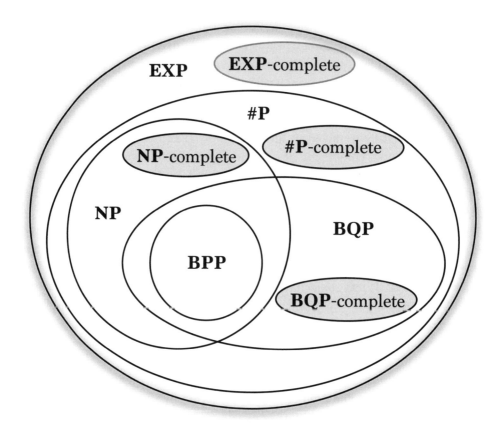

Figure 1.1 Believed relationships between the complexity classes most relevant to quantum computing. **BPP** is the class of polynomial-time probabilistic classical algorithms. **NP** is the class of problems verifiable in polynomial time using classical algorithms. **NP**-complete are the subset of **NP** problems polynomial-time reducible to any other problem in **NP**, similarly for other 'complete' problems. **BQP** is the class of probabilistic algorithms solvable in polynomial time on universal quantum computers. **#P** is the set of counting problems that count satisfying solutions to **P** problems (**P** is the same as **BPP** but deterministic rather than probabilisitic). **EXP** is the class of all algorithms that require exponential time. Note that it is actually unproven whether $\mathbf{P} = \mathbf{BPP}$ or $\mathbf{P} \subset \mathbf{BPP}$. There are examples where the best known **BPP** algorithms outperform the best known **P** algorithms, which could arise because the two classes are inequivalent or because we simply have not tried hard enough to find the best deterministic algorithms. Furthermore, though it is known that $\mathbf{P} \subseteq \mathbf{NP}$, it is not known whether $\mathbf{BPP} \subseteq \mathbf{NP}$. For the sake of illustration in our Venn diagram we have taken the view that it is. **BPP** is regarded as the class of problems efficiently solvable on universal Turing machines (i.e., classical computers), whereas **BQP** is the class efficiently solvable on universal quantum computers. The computational superiority of quantum computers is based on the (strongly believed, yet unproven) assumption that $\mathbf{BPP} \subset \mathbf{BQP}$.

The first demonstrations of digital computer networks were nothing more than simple two-party, point-to-point (P2P) communication. However, the internet we have today extends far beyond this, allowing essentially arbitrary worldwide networking across completely ad hoc networks comprising many different mediums, with any number of parties, in an entirely plug-and-play and decentralised fashion. Similarly, elementary demonstrations of quantum communication have been performed across a small number of parties, and much work has been done on analysing quantum channel capacities in this context. But, as with digital computing, demand for a future *quantum internet* is foreseeable, enabling the arbitrary communication of quantum resources, between any number of parties, over ad hoc networks.

The digital internet may be considered a technology stack, such as TCP/IP (Transmission Control Protocol/Internet Protocol), comprising different levels of abstraction of digital information [177]. At the lowest level we have raw digital data we wish to communicate across a physical medium. Above this, we decompose the data into packets. The packets are transmitted over a network, and TCP is responsible for routing the packets to their destination and guaranteeing data integrity and Quality of Service (QoS). Finally, the packets received by the recipient are combined and the raw data are reconstructed.

The TCP layer remains largely transparent to the end-user, enabling virtual software interfaces to remote digital assets that behave as though they were local. This allows high-level services such as the File Transfer Protocol (FTP), the worldwide web, video and audio streaming and outsourced computation on supercomputers, as though everything were taking place locally, with the end-user oblivious to the underlying networking protocols, which have been abstracted away. To the user, YouTube videos or Spotify tracks behave as though they were held as local copies. And FTP or DropBox allows storage on a distant data centre to be mounted as though it were a local volume. We foresee a demand for these same criteria in the quantum era.

In the context of a quantum internet, packets of data will instead be quantum states, and the transmission control protocol is responsible for guiding them to their destination and ensuring quality control.

Our treatment of quantum networks will be optics heavy, based on the reasonable assumption that communications channels will almost certainly be optical, albeit with many possible choices of optical states and mediums. However, this does not preclude nonoptical systems from representing quantum information that is not in transit, and we consider such 'hybrid' architectures in detail, as well as the interfacing between optical and nonoptical systems. Indeed, it is almost certain that future large-scale quantum computers will not be all-optical, necessitating interfacing different physical architectures.

Shared quantum entanglement is a primitive resource with direct applications in countless protocols. This warrants special treatment of quantum networks that do not implement a full network stack but instead specialise in just this one task – entanglement distribution. We will see that such a specialised network will already be immensely useful for a broad range of applications, and its simplicity brings with it many inherent advantages.

The quantum internet will enable advances in the large-scale deployment of quantum technologies. Most notable, in the context of quantum computing it will allow initially very expensive technology to be economically viable and broadly accessible via the outsourcing of computations for both consumers who cannot afford quantum computers and to well-resourced hosts who can – *cloud quantum computing*.

With the addition of recent advances in homomorphic encryption and blind quantum computing, such cloud quantum computing can be performed securely, guaranteeing privacy of both data and algorithms, secure even against the host performing the computation. This opens up entirely new economic models and applications for the licensing of compute time on future quantum computers in the cloud.

The unique behaviour of quantum computing, in terms of the superclassical scaling in its computational power, brings with it many important economic and strategic considerations that are extremely important to give attention to in the postclassical world.

But quantum technologies extend far beyond computation. Many other exciting applications for controlled quantum systems exist, with new ones frequently emerging. Thus, the quantum internet will find utility beyond cloud quantum computing, enabling the global exchange of quantum resources and assets. This could include the networking of elementary quantum resources such as state preparation, entanglement sharing and teleportation and quantum measurements or scale all the way up to massively distributed quantum computation or a global quantum cryptography network.

It is hard to foresee the future trajectory of quantum technology, much as no one foresaw the advances digital technology has made over the last half century. But it is certain that as the internet transformed digital technology, the quantum internet will define the future of quantum technologies.

Part I
Classical Networks

2

Mathematical Representation of Networks

We begin by turning our attention to defining a mathematical construction for the representation of (quantum and/or classical) networks, which we will subsequently rely on heavily in our framework for quantum networks. This encompasses representing networks as graphs, representing the cost of communications within the network and how to optimise network routing to minimise costs. These notions will be essential in our treatment of quantum networks.

2.1 Graph-Theoretic Representation

We consider a classical network to be a weighted, directed graph,

$$G = (V, E), \tag{2.1}$$

where vertices represent *nodes* ($v \in V$) in the network and the weighted edges represent communication *links* ($e \in E$) between neighbouring nodes.

A node could be, for example, data storage, a classical computer implementing a computation, a router that switches the connections between incoming and outgoing links or an end-user – anything that communicates with the network, sender or receiver. A link, on the other hand, is any arbitrary means of communication between nodes, such as optical fibre, satellite, radio, electrical, smoke signals, tin cans connected by a taut piece of string or a well-trained carrier pigeon. In the protocols to be described here, it is completely irrelevant what the specific mediums for communication are. Rather what matters are *costs* quantifying the relative performance of different links.

A key feature of the global internet is redundancy. In a packet-switched environment, when sending identical packets twice might each follow entirely different routes to their common destination. Node-to-node redundancy is easily accommodated for in the graph-theoretic model by allowing multiple distinct edges between nodes. It is extremely important to accommodate multiple edges in network graphs, because redundant routes provide a direct means by which to load-balance a route. So, for example, a hub in Australia might connect to a sister hub in New Zealand using both a fibre-optic undersea cable and simultaneously via a satellite uplink. If the faster of the two connections is running out of capacity, a proportion of the packets can simply be switched to the other link, thereby balancing the load. For this reason we abstain from using an adjacency matrix representation for network graphs, because they do not accommodate redundancy.

2.2 Cost Vector Analysis

The edge weights in G represent the *costs* $(\vec{c})$ associated with using that link.

Definition 1 (Network cost metrics) *Cost metrics satisfy the following properties:*

- *Identity operations: If a channel performs nothing, its associated cost is zero,* $c(\hat{I}) = 0$.
- *Triangle inequality:*
  $c(v_1 \rightarrow v_2 \rightarrow v_3) \leq c(v_1 \rightarrow v_2) + c(v_2 \rightarrow v_3),$
  across all paths $v_1 \rightarrow v_2 \rightarrow v_3$. *In the case of strict equality under addition we refer to the cost as a strictly additive cost.*
- *Positivity:* $c \geq 0$. *This ensures that shortest-path algorithms will function correctly. It is also congruent with the intuitive expectation that data travers- ing a communications channel are not somehow better off than if they had not traversed that channel at all.*

The reason we demand that costs have a distance interpretation is so that graph-theoretic pathfinding algorithms are applicable, allowing us to build upon the vast preexisting understanding of graph theory. Ideally we would like equality in costs' triangle inequality, which yields an exact cost. But often this is not possible and we are satisfied with the inequality, which simply dictates an upper bound on cost.

A detailed discussion of some of the major costs that realistic quantum networks will be subject to is presented in Chapter 8.

A *route* between two nodes, Alice (A) and Bob (B), of the network, G, is an acyclic subgraph connecting those nodes, $R_{A \rightarrow B} \subseteq G$. In general ad hoc networks there will typically be multiple paths between two nodes $A \rightarrow B$. For a particular cost metric, the cost of an entire route is simply the sum of the costs of each of the constituent links,

Definition 2 (Route costs) *The net cost of a route* $A \rightarrow B$, *using cost metric* $c(A \rightarrow B)$, *traversing nodes* v_i *is*

$$c(R_{A\rightarrow B}) = \sum_{i=1}^{|R_{A\rightarrow B}|-1} c(v_i \rightarrow v_{i+1}), \qquad (2.2)$$

where v_i *is the ith node in the route* $R_{A\rightarrow B}$.

Figure 2.1 illustrates a simple example network with all of its available routes, $R_{A\rightarrow B} \subseteq G$. Figure 2.2 illustrates the optimal path for $A \rightarrow B$ based on edge weights.

In a given network, it is unlikely that only a single cost metric will be of interest when determining optimal routings. There may be a trade-off between different measures. For example, for time-critical applications the cost of a route might be considered a combination of both dollar cost and latency – a satellite has very low latency but is extremely expensive, whereas a carrier pigeon is slow but cheap (and prohibited by PETA). What is the best trade-off between the two?

To accommodate this, we allow the *net cost* of a route to be defined as an arbitrary function of other primitive costs of the route,

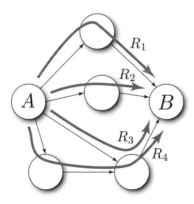

Figure 2.1 Example of a simple network with multiple routes $A \rightarrow B$. Note that R_3 and R_4 are competing with one another for use of the last link, which the routing strategy, $\mathcal{S}$, will need to resolve if multiple simultaneous transmissions are taking place.

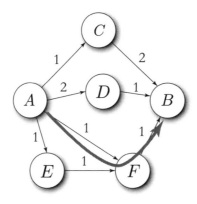

Figure 2.2 The same network graph from Figure 2.1, with links weighted by some arbitrary cost metric. Applying a shortest-path algorithm yields the optimal route between Alice and Bob to be $A \rightarrow F \rightarrow B$, which incurs a net cost of $c = 2$, as opposed to all other routes, which incur a net cost of $c = 3$.

Definition 3 (Net routing cost) *The net cost of a route $A \rightarrow B$ is given by*

$$c_{\text{net}}(R) = f_{\text{cost}}(\vec{c}(R)), \tag{2.3}$$

where c_{net} is a single numeric value representing the net cost as calculated from an arbitrary cost function, f_{cost}, of the vector of associated costs.

Note that the net routing cost need not be a metric, because the cost function could be arbitrary. The net cost can be thought of as a ranking for routes but not necessarily as a metric that accumulates across routes, because it already captures all of these accumulations.

Equation (2.3) gives us the net cost of a given route. For multiple users we would like to simultaneously optimise the cost across all users of the network. Thus, we define the routing cost for the entire network to be the following.

Definition 4 (Network routing cost) *The net routing cost of all costs over all active routes $\vec{R}$ is*

$$c_{\text{total}}(\vec{R}_{\vec{A} \rightarrow \vec{B}}) = \sum_{r \in \vec{R}_{\vec{A} \rightarrow \vec{B}}} c_{\text{net}}(r), \tag{2.4}$$

where $\vec{R}_{\vec{A} \rightarrow \vec{B}}$ is a set of active routes connecting each pair $A_i \rightarrow B_i \; \forall \, i$.

2.3 Routing Strategies

A *strategy*, $\mathcal{S}$, is simply an algorithm that chooses a route based on the starting and finishing nodes of a communication and also updates the cost vectors within the network associated with the utilisation of that route.

Definition 5 (Routing strategies) *A routing strategy is defined by*

$$\mathcal{S}(i, j, \vec{c}) \rightarrow \{k, \vec{c}\,'\},$$

$$i, j \in V,$$

$$k \in \{R_{v_i \rightarrow v_j}\}, \tag{2.5}$$

where $\mathcal{S}$ denotes the strategy, k is a route, i and j are the source and destination nodes of the route and $\vec{c}$ is a vector of associated costs.

The goal of the strategy $\mathcal{S}$ is to minimise a chosen cost measure.

No particular route through a network is going to have infinite capacity and therefore we cannot typically always reemploy the same most cost-effective route for all data. Particularly in multi-user networks, as routes are employed for communicating quantum states, their cost metrics may change according to load or other external influences. Alternatively, some routes may come into and out of operation. For example, a satellite requiring line-of-sight communication may oscillate in and out of sight, thereby periodically enabling and disabling respective network routes. For this reason, it is important that strategies accommodate dynamic changes in the network. This is easily accounted for by letting the edge weights in our network graph be a function of time, G_t, which are updated via the application of a strategy that may also be time dependent.

Definition 6 (Time-dependent routing strategies) *A time-dependent strategy, $\mathcal{S}_t$, updates the network graph, G_t, at each time step t,*

$$G_{t+1} = \mathcal{S}_t(G_t). \tag{2.6}$$

*$\mathcal{S}_t$ could be any **BPP** algorithm, deterministic or probabilistic.*

For example, the network might have bandwidth restrictions on some links, in which case if more than a certain amount of data is transmitted through a link it is no longer available for use until previous transmissions have completed. Or, based on market dynamics, the dollar cost of utilising a link may change with its demand.

This type of cost minimisation approach to routing is analogous to *distance-vector routing protocols* in classical networking theory.

2.4 Strategy Optimisation

Clearly the goal when choosing routing strategies is to minimise the total cost, Eq. (2.3). That is, solving the following optimisation problem.

Definition 7 (Strategy optimisation) *The optimisation of strategies with a network comprising net costs c_{total} is given by*

$$c_{\text{min}} = \min_{S}(c_{\text{total}}),$$

$$S_{\text{opt}} = \underset{S}{\text{argmin}}(c_{\text{total}}). \tag{2.7}$$

Choosing optimal strategies is a challenging problem, potentially requiring complex, computationally inefficient optimisation techniques. Strategy optimisation is an example of resource allocation whose optimal solutions are often notoriously difficult to solve exactly, residing in complexity classes like **NP**-complete (or worse!). In general, the number of possible routes through a graph will grow exponentially with the number of vertices. Thus, explicitly enumerating each possible route is generally prohibitive for large networks, unless some known structure provides 'shortcuts' to optimisation. Having said this, Dijkstra's shortest-path algorithm is the perfect counterexample, demonstrating that although an exponential number of routes may exist between two points, an optimal one can be found in **P**.

Ad hoc Operation vs Central Authorities

When considering strategy optimisation, the first question to ask is 'Who performs the optimisation, and who has access to what information?'

In terms of who performs the optimisation, the two main options are that either each node is responsible for optimising the routes of packets passing through it (INDIVIDUAL algorithms) or there is a reliable and trusted central mediating authority who oversees network operation and performs all strategy decision making (CENTRAL algorithms).

In the case of INDIVIDUAL algorithms, the required knowledge of the state of the network could be obtained using network exploration algorithms or gateway protocols.

On the other hand, for CENTRAL algorithms, either network exploration could be employed or, alternatively, the network policy could require nodes to notify the central authority upon joining or leaving the network. The former introduces an overhead in classical networking resource usage, because network exploration must be performed routinely to keep the ledger of nodes up to date. The latter, on the other hand, avoids this but introduces a point of failure, in that all network participants must be reliable in notifying the central authority as required by the network policy. Failure to do so could result in invalid or suboptimal strategies.

Local vs Global Optimisation

There are two general approaches one might consider when choosing strategies: *local optimisation* (LOCAL) and *global optimisation* (GLOBAL). LOCAL simply takes each state to be communicated, one by one, and allows it to individually choose an optimal routing strategy based on the state of the network at that moment. GLOBAL is far more sophisticated and simultaneously optimises the sum of the routing costs, Eq. (2.4), of all currently in-demand routes.

To implement LOCAL optimisation, either INDIVIDUAL or CENTRAL algorithms may be employed. On the other hand, GLOBAL optimisation necessarily requires a CENTRAL algorithm, because it requires knowledge of the entire state of the network, which is collectively optimised.

Because GLOBAL represents the class of all algorithms that take all network costs by all packets into consideration, it must clearly perform at least as well as LOCAL, which only takes into consideration the costs of a given packet. But we expect GLOBAL to perform better than LOCAL in general, owing to the additional information it takes into consideration. We express this as LOCAL$\subset$GLOBAL. However, GLOBAL requires solving a complex, simultaneous optimisation problem, which is likely to be computationally hard, whereas LOCAL can be efficiently solved using multiple independent applications of, for example, an efficient shortest-path algorithm (so-called GREEDY algorithms).

A further stumbling block for GLOBAL is that it requires some central authority, responsible for the global decision making, to have complete, real-time knowledge of the state of the entire network. This may be plausible for small local area networks but would clearly be completely implausible for the internet as a whole. So it is to be expected that different layers and subnets in the network hierarchy will employ entirely different strategy optimisation protocols. This is certainly reminiscent of the structure of the present-day internet.

Roughly speaking, we might intuitively guess that at lower levels in the network hierarchy, responsible for smaller subnets, there will be a tendency towards the adoption of GLOBAL strategies, as full knowledge of the state of the subnet is readily obtained and maintained. However, as we move to the highest levels of the network hierarchy (e.g., routing of data across international or intercontinental boundaries), we might expect more laissez-faire (i.e., GREEDY) strategies to be adopted, because the prospects of enforcing a central authority with full knowledge of the state of the internet, who is also trusted by all nations to fairly and impartially allocate network resources and mediate traffic, are highly questionable.

We will not aim to comprehensively characterise the computational complexity of GLOBAL strategies. However, in Chapter 9 we will present some elementary analyses of several toy models for realistic strategies. Some such strategies are efficient and, although not optimal, nonetheless satisfy certain criteria we might expect.

Future developments in the optimisation techniques required for GLOBAL strategies may improve network performance, leaving our techniques qualitatively unchanged.

When employing LOCAL, on the other hand, things are often far simpler. If we are optimising over a cost metric satisfying the distance interpretation, we may simply employ a shortest-path algorithm to find optimal routes through the network.

If one were to become even more sophisticated, one might even envisage treating network resource allocation in a game theoretic context, which we will not even begin to delve into here.

3

Network Topologies

Because quantum (or classical) networks inherently reside on graphs, it is important to introduce some of the key graph structures of relevance to networking and some of their properties of relevance to quantum networking protocols.

Let the graph G representing the network be

$$G = (V, E), \tag{3.1}$$

with vertices V and edges E. In principle a network could be characterised by any connected graph whatsoever. However, there are certain structures and patterns that emerge very frequently and deserve special attention.

It is paramount that quantum networking protocols have the capacity to deal with the diverse network topologies that are likely to present themselves in the future real-world quantum internet. Some of the graph-theoretic algorithms that we rely on are computationally efficient for *arbitrary* graph topologies, even more so for certain classes of graphs exhibiting particular structure, such as tree graphs or complete graphs. Others, however, are computationally inefficient in general but may have efficient approximation algorithms for some or all classes of topologies.

We will now review some of the graph structures most likely to arise in quantum networks, learning from the structures that have become ubiquitous in classical networking. Understanding the basic mathematical properties of these different network topologies is extremely important to take into consideration when designing future quantum networks, because they strongly impact important features such as construction cost of the network infrastructure, routing cost vector analysis, likelihood of successful routing and transmission time.

A summary of the basic mathematical characteristics of the topologies presented is shown in Table 3.1, specifically showing the number of edges and vertices and *diameter* of the topologies (i.e., the distance between extremal points in the network).

Table 3.1 *Summary of the mathematical characteristics of different network topologies.*

Topology	Vertices ($|V|$)	Edges ($|E|$)	Diameter (d)
Point-to-point	2	1	1
Linear	$|V|$	$O(|V|)$	$|V|$
Complete	$|V|$	$O(|V|^2)$	1
Lattice	mn	$O(mn)$	$O(m+n)$
Tree	$|V|$	$O(|V|)$	$O(\log|V|)$
Percolation	$p_{vertex} \cdot |V|$	$p_{edge} \cdot |E|$	variable
Random	$p_{vertex} \cdot |V|$	$p_{edge} \cdot O(|V|^2)$	variable
Scale-free	$|V|$	$|E|$	$O(\log\log|V|)$

3.1 Point-to-Point

The most trivial network topology, which also acts as the elementary primitive from which our other topologies will be constructed, is a simple dedicated point-to-point (P2P) connection between two parties, where the sender and recipient of a packet reside on neighbouring nodes.

Such P2P connections may be reserved exclusively for the two connected neighbouring nodes. In this instance, the packets' ROUTING QUEUES trivially specify just the recipient. Alternately, the P2P link may be an intermediate step between more distant sender–recipient pairs.

In the case whereby the P2P connection is reserved exclusively for a particular sender–recipient pair, the link has the property that there is no competition between multiple users sharing the channel, and the quantum network stack need not concern itself with dynamic routing strategies.[1] This significantly simplifies network scheduling algorithms, and a FIRST-COME FIRST-SERVED (i.e., chronologically ordered FIFO [first in, first out] queue) strategy may be employed. Furthermore, packet collisions cannot occur, thereby improving network efficiency.

In the case whereby the P2P connection is not reserved for exclusive use between a single sender–recipient pair but is shared between different competing routes in the network, the importance of network routing strategies manifests itself. Now competition for access to the channel will reduce network efficiency, scaling inversely against the number of network participants, and the priorities and costs of packets must be tallied for the purpose of implementing routing strategies.

---

[1] Assuming that the P2P channel has sufficient capacity to meet demand and exhibits better cost metrics than other potential redundant, indirect routes.

3.2 Linear

A linear graph topology has very simple properties. The number of edges simply scales as

$$|E| = |V| - 1, \tag{3.2}$$

and the graph diameter is simply the number of vertices,

$$d = |V|. \tag{3.3}$$

There are limited routing considerations for such a topology because there is always exactly one route between two points, although buffering issues may still arise under congestion.

Because there is no path redundancy, linear graphs are vulnerable to node failures, because the failure of a single node disconnects the network.

3.3 Complete

The complete graph, denoted $K_{|V|}$, is a $|V|$-vertex graph where every vertex has an undirected link to every other. From a networking point of view, this can be regarded as the extremity of exclusive-use P2P networking, whereby every node has a direct link with every other. Thus, any sender can directly communicate with any receiver, via a dedicated direct channel, with no need to utilise any indirect routes. This topology has the favourable property that although any node can communicate with any other, by exclusively utilising direct P2P links we achieve several benefits:

- Competition for the use of links can be eliminated, minimising congestion and the need for buffering (i.e., quantum memory).
- Network costs can typically be minimised, because every route only traverses a single link and there will be no accumulation of costs.
- The network has maximal route redundancy, making it the most tolerant against link failures.[2]
- A trivial FIRST-COME FIRST-SERVED routing strategy can be employed, eliminating the need for any dynamic or computationally complex strategies.

---

[2] To disconnect a given node v from the network, all $|v| - 1$ links emanating from it must be broken; otherwise, redundant routes to the remainder of the network will exist.

- If the network allows indirect routes to be established, the maximal redundancy of the topology also maximises the ability for routing strategies to engage in load balancing across routes.
- In the special case of a symmetric complete graph, whereby all edge weights are approximately equal, the shortest path between any two nodes is trivially the P2P link between them, and no complex scheduling algorithms are required.

However, these highly desirable benefits come at the expense of requiring the most elaborate and expensive network, with maximal interconnectedness.

This type of topology could arise in, for example, international-scale networks, where links of very high bandwidth (and value) between nations or continents need to be maximally utilised, which would be undermined by sparse, shared network topologies. Additionally, in this instance route redundancy will be highly valued, because the isolation of one continent from another would be catastrophic to the functioning of the global network.

The number of edges scales as

$$|E| = O(n^2). \tag{3.4}$$

Clearly route finding is trivial, because there is always a direct P2P link from sender to receiver, with no possibility of collisions with other packets, requiring $O(1)$ search time (assuming that all users are communicating only via their direct links with one another, which may not strictly be the case when costs are factored into strategies).

3.4 Lattice

A lattice graph is simply an $n \times m$ lattice of vertices (of any geometry; e.g., squares), connecting each vertex to its immediate geometric neighbours. The number of edges scales obviously as

$$|E| = O(mn). \tag{3.5}$$

This type of graph is useful when link costs are measured in terms of Euclidean distances and nodes have nearest neighbour links.

A slightly distorted lattice graph, in which vertices have been dragged around geometrically to match, for example, cities within a country, closely resembles the topology of the network. Similarly, if the nodes represent houses in the street layout of a highly regular city like Manhattan, a lattice may be a good approximation.

In the case of a balanced lattice, in which all edges are of equal weight, the cost of a route is the sum of the number of steps in the vertical and horizontal directions, also known as the Manhattan or L_1 distance,

$$L_1 = |x_{\text{start}} - x_{\text{finish}}| + |y_{\text{start}} - y_{\text{finish}}|. \tag{3.6}$$

In this case, route finding is simplified, because *all* routes, which strictly traverse in one direction vertically and one direction horizontally, are optimal and of equal distance. Thus, the diameter (maximum number of hops between any two points) on the network is

$$d = O(m + n). \tag{3.7}$$

3.5 Tree

A tree is a graph containing no cycles, only *branches*. There are many uses for tree graphs, but one property is of particular convenience in many applications: because the graph is acyclic, there is always exactly one path from any vertex to any other. This mitigates the need for shortest-path algorithms designed for general graphs and simplifies route finding algorithms (to be discussed in Section 4.1). However, this brings with it the drawback that the topology is most vulnerable to link failures, because the removal of any link from the tree will separate it into a multipartite graph, making communication between the disjoint subgraphs (which are also trees) impossible, because there are no redundant routes. In a sense, tree graphs can be considered the polar opposites of complete graphs.

Trees are specified entirely by *branching parameters* (b_i); i.e., the number of child nodes emanating from a given node, i. In general, branching parameters may be distinct for each node, although often trees with symmetries in their branching structures are considered, such as the balanced trees discussed in Section 3.5. A node terminates a branch if its branching parameter is zero (i.e., it has no children).

The *depth* (d) of a tree is the maximum number of steps from the root node to a terminating node with no children. The depth scales between $d = O(|V|)$ for the trivial linear tree ($b_i = 1$) and $d = O(\log |V|)$ for nontrivial branching parameters ($b_i \neq 1$).

The worst-case number of edges that must be traversed to reach any vertex from any other is

$$O(\log |V|), \tag{3.8}$$

known as the *diameter* of the graph, which implies that accumulated cost metrics scale similarly. Trees are the most frugal graphs in their number of edges, which are fixed at

$$|E| = |V| - 1, \tag{3.9}$$

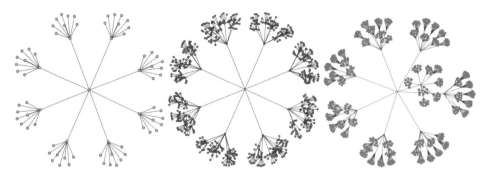

Figure 3.1 Balanced tree graphs with branching factor $b = 8$ and depths $d =$ 3, 4, 5. Despite having no redundant paths, the hierarchical structure of balanced trees somewhat resembles that of real-world networks, which are typically decomposed into a pyramid scheme of progressively smaller subnetworks.

irrespective of the branching parameters; i.e., because the graph is strictly acyclic, every addition of an edge requires the addition of exactly a single vertex. This makes tree graphs the cheapest to construct in terms of physical resource usage.

Balanced Tree

A balanced tree is a tree with a regular, self-similar structure, in which every node at a given depth is the parent of the same number of subnodes, all separated by the same edge weights. That is, the network has a hierarchical structure, subdivided into identically structured subnetworks. Such a network is characterised by just two parameters: the branching parameter, b, and the depth, d. Some examples of balanced trees with different b and d are shown in Figure 3.1.

This type of structure is (approximately) natural in many realistic scenarios. Consider, for example, a network containing a hierarchy of clusters of nodes representing a local area network, followed by a neighbouring internet router, followed by a citywide router, followed by a countrywide router. In such a case, this type of general structure is typical (although more realistically one might expect the branching parameter to vary with depth).

A special case is when $d = 1$, which we refer to as a *star* graph. This might arise naturally when a series of subnets are connected together via a central router, with no further hierarchy in the network.

Random Tree

Though balanced trees accurately capture the hierarchical nature of realistic networks, they are somewhat contrived in their perfect symmetry. The subnetworks in

Figure 3.2 Random trees with different randomised branching parameters (higher *b* on the right). When a node has zero branches, it terminates the branch. This type of graph topology qualitatively captures the hierarchical, yet ad hoc qualities of many real-world networks and may act as a useful test model for simulations.

a given network are not likely to actually all be identical. Random trees are perhaps more realistic, in that their tree structure captures the hierarchical nature of real-world networks and also their highly ad hoc nature.

To construct a random tree we simply randomly choose a branching parameter, according to some arbitrary distribution, for every node. When a node has $b_i = 0$, it terminates the lineage. Some examples of random trees are shown in Figure 3.2.

Minimum Spanning Tree

A *spanning tree* S, of a graph G, is a tree subgraph $S \subset G$ containing every vertex of G. The *weight* of a spanning tree is the sum of all its constituent edge weights. Thus, the *minimum spanning tree* (MST) is a spanning tree that minimises net weight. An example is shown in Figure 3.3.

The calculation of MSTs is most likely to come into consideration when actually performing the initial construction of networks, where we wish to connect all nodes in the network, but using the most frugal possible physical resources. MSTs serve this purpose, and because they are trees, they inherit all of the same properties of tree networks.

In general, the MST of a graph is not unique, and there may be an arbitrarily large number of completely differently structured MSTs all with the same minimum weight.

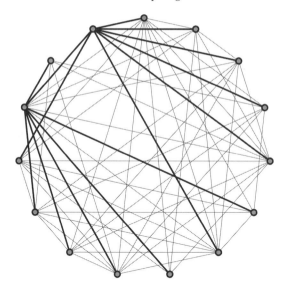

Figure 3.3 A random graph with an MST highlighted.

3.6 Percolation

A variation on any graph is to instead have a randomised implementation of it, whereby each of the possible edges or vertices occur with some probability, p_{edge} or p_{vertex}, otherwise deleted. These are referred to as *edge percolation* and *site percolation* graphs, respectively.

For any given graph, its associated percolation graph has average vertex and edge counts

$$|E|_{\text{av}} = p_{\text{edge}} \cdot |E|,$$
$$|V|_{\text{av}} = p_{\text{vertex}} \cdot |V|. \tag{3.10}$$

Adjusting $p_{\text{edge/vertex}}$ allows us to tune between the desired graph G (when $p_{\text{edge/vertex}} = 1$) and the completely disconnected graph (when $p_{\text{edge/vertex}} = 0$).

This model is very useful in real-world applications, allowing unreliable channels/nodes to be incorporated into our network model. The analysis of such percolation networks is invaluable for understanding the robustness of such networks to channel and node failures.

Note that percolation graphs might be disjoint with sufficient defects, in which case the respective network becomes unreliable. Specifically, with sufficiently low $p_{\text{edge/vertex}}$, 'islands' may form in the network topology; i.e., small segregated networks that are unable to interface with the remainder of the network.

For asymptotically large percolation graphs, *percolation theory* provides thresholds for $p_{\text{edge/vertex}}$ such that routes across the network exist in asymptotic limits.

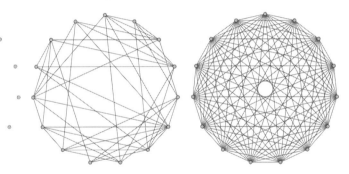

Figure 3.4 The 15-vertex random graph, where edges are present with probabilities $p_{\text{edge}} = 0, 0.5, 1$ (left to right).

3.7 Random

We refer to a random graph as being one in which edges between each pair of vertices occur with some probability p_{edge}. No vertices are removed from the network, although some may have order $|v| = 0$; i.e., $p_{\text{vertex}} = 1$. This can be thought of as the edge percolation graph of the complete graph $K_{|V|}$.

The average number of edges in such a network scales as

$$|E|_{\text{av}} = p_{\text{edge}} \cdot O(|V|^2). \tag{3.11}$$

Some examples are shown in Figure 3.4.

3.8 Hybrid

Real networks are highly unlikely to fit the exact form factor of any of the classes of graphs presented above. Rather, a truly global internet is inevitably going to comprise many subnetworks, each structured completely independent of one another, with little consistency or large-scale planning between them. Who thinks about the broader structure of the global internet when setting up their office network?

For example, at the global scale, it is entirely plausible that the internet might take on a random tree-like structure. But when we get down to a lower level, the tree structure vanishes and is replaced by all manner of different network topologies, run and maintained by different organisations in their own distinct ways.

Furthermore, the real-world internet is not simply a hierarchy of different types of well-known graph structures. Rather, it takes the form of 'glued' graphs, whereby networks running over different mediums, or via different operators, each exhibit their own independent graph topologies, meeting at interconnect points that join the different networks. Typically this yields redundancy in the routes between different nodes, ushering in the need for combinatorial optimisation techniques when allocating network resources.

This hybrid network topology is the norm today in our classical internet, and it is entirely foreseeable that a similar trend will emerge in the future quantum internet as quantum technologies become more mainstream and their networking less well structured and competing, redundant links are in place.

3.9 Network Robustness

A key feature of any network topology is its robustness against node or channel failures. This is important from the perspective of naturally occurring hardware faults and also from a geostrategic perspective, where adversaries may be launching attacks against the network. In general, there are two main contributing factors to network robustness:

- Redundancy: the number of redundant paths between two points in a graph stipulates how many backups there are to finding a route to a destination in the advent of one route failing.
- Diameter: the chance of a data packet encountering a faulty node/channel increases with the number of hops required to reach its destination. Graphs with smaller diameter are hence less vulnerable.

The extreme case of network robustness is the complete graph, K_n, which has P2P links between every pair of nodes. Therefore, if a single channel fails, there are *always* alternate paths taking us between nodes. On the opposing extreme are tree graphs, which contain no redundancy whatsoever and just a single failure will disconnect the network, making certain routes impossible. Scale-free networks sit in the intermediate zone and are relatively robust against the failure of random nodes/links but are vulnerable to conspiratorial failures, which target the elite, highly connected hub-nodes.[3]

Figure 3.5 illustrates some examples of the robustness of these two extreme cases to link and node failure.

---

[3] The 1%.

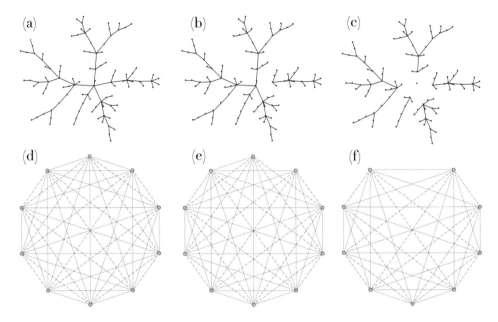

Figure 3.5 Robustness of network topologies to node and link deletion. Examples of (a) a tree graph and (d) a complete graph, K_n, with $n = 10$. (b), (e) The same graphs subject to a single link failure. (b) The failure disconnects the tree graph into a bipartite graph, whereas (e) the complete graph's connectivity is unhindered as alternate routes exist between all nodes. (c) A single node failure disconnects the tree graph into a $|v|$-partite graph, where $|v|$ is the order of the vertex at which failure occurs. (f) The complete graph, on the other hand, is simply reduced to a K_{n-1} graph, with no loss of connectivity. Thus, tree graphs are the most vulnerable network topologies to node/link failures, whereas complete graphs are the most robust.

4

Network Algorithms

Having introduced some of the more relevant graph structures, we now introduce some of the key graph-theoretic algorithms of direct relevance to networking theory [48]. In graph theory, many fundamental problems are believed to be computationally hard to solve, often **NP**-complete. However, there are several important graph algorithms that are (very) classically efficient to solve and that are of great utility to us as network architects.

4.1 Network Exploration and Pathfinding

Here the goal is to systematically explore every vertex in an unknown graph exactly once to reconstruct the entire network graph or to find a target node with unknown location (which can obviously be achieved if the former can be). The two main approaches are *breadth-first-search* (BFS) and *depth-first-search* (DFS) algorithms. In both cases we begin at a starting (root) node, from which we wish to explore the entire graph by only following edges to nearest neighbours one at a time.

In BFS we proceed from the root node to visit every one of its neighbours. Having done so, and created a list of those neighbours, we proceed to the neighbours of the neighbours, and so on, until every vertex in the graph has been visited or the target node has been found.

In DFS, on the other hand, we begin by following a single arbitrary path until we reach a dead-end, at which point we backtrack until we reach a branch leading to a vertex we had not previously visited.

Examples of these two algorithms are shown in Figure 4.1.

Both BFS and DFS guarantee visiting every vertex in a connected graph and do so using only nearest neighbour transitions. Such algorithms are therefore very useful for network discovery.

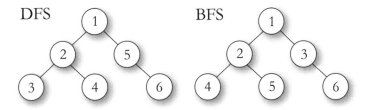

Figure 4.1 Comparison of the order in which vertices are explored, using the breadth-first-search (BFS) and depth-first-search (DFS) algorithms, where vertex 1 is the root vertex.

The BFS algorithm is particularly applicable to pathfinding in ad hoc networks. Consider the situation where there is no central authority with full knowledge of the network overseeing network operation. Rather, everyone needs to figure things out for themselves by only interrogating their neighbours, to whom they have direct connections. This directly leads to a BFS algorithm, where a node speaks to each of its neighbours in turn, who subsequently do the same thing, yielding a recursive algorithm. This can be naturally parallelised, because each node can be interrogating its neighbours independently, thereby implementing a distributed BFS algorithm. Note that, when searching for a target node, though the BFS algorithm obviously finds the target using the smallest number of hops (i.e., a lowest-order route), it need not necessarily find the route with the lowest cost (which is distinct from the number of hops in general). Shortest-path algorithms require *a priori* knowledge of the full network graph, discussed in Section 4.2.

Both BFS and DFS exhibit runtime

$$O(|V| + |E|),\qquad(4.1)$$

where $|V|$ and $|E|$ are the number of vertices and edges, respectively. Thus, these graph exploration algorithms reside in the complexity class **P** and are classically efficient.

4.2 Shortest Path

In graph theory, the shortest-path problem is that of finding a subgraph of a given graph G, connecting two vertices, $A \rightarrow B \subset G$, such that the sum of its edge weights is minimised. In the context of our application to route-finding, this amounts to finding a route that minimises cost.

The first proven shortest-path algorithm was invented by and named after Dijkstra [55], which requires runtime

$$O(|V|^2),\qquad(4.2)$$

which has since been improved to

$$O(|E| + |V| \log |V|) \tag{4.3}$$

by [70] using min-priority queues implemented via Fibonacci heaps. Thus, the shortest-path algorithm resides in **P** – one of the relatively few, and highly valuable, optimisation problems that is classically efficient. Subsequently, a number of improvements and variations on Dijkstra's algorithm have been proposed, most notably the A^* algorithm [85], which has found widespread modern use, using a heuristic approach to improve performance over Dijkstra.

Formally, let $\vec{R}$ be the set of all routes $A \rightarrow B$. Then,

$$c_{\text{opt}} = \min_{r \in R} \left(\sum_{i \in r} c_{\text{net}}(i) \right). \tag{4.4}$$

To implement the shortest-path algorithms discussed above, the party performing the calculation requires knowledge of the full network graph. In an ad hoc network, where users might be added to or removed from the network arbitrarily, this is not necessarily the case.

One solution is for a central authority to be responsible for maintaining a ledger of all network participants and their connectivity, which users are required to notify upon joining or leaving the network. The central authority may then apply shortest-path calculations, which may be queried by users. However, a disruption in connection to the central authority, or failure of nodes to notify the central authority upon joining or leaving the network, introduces a point of failure into the operation of the protocol.

Another approach, which does not require a reliable central authority, is for users to implement network exploration algorithms each time they wish to perform a shortest-path calculation. This facilitates truly ad hoc networking but incurs the cost overhead associated with nodes frequently implementing network exploration. However, network exploration is a purely classical algorithm that may run entirely over the classical network and therefore incurs no cost in quantum resources.

With this approach, a new node can join the network without having to know anything about the topology of the network. Similarly, upon leaving the network, it need not notify anyone, because a future interrogation by a neighbour will be detected as a nonexistent node. The BFS is therefore highly suited to ad hoc operation. In fact, present-day internet gateway protocols essentially implement a distributed version of BFS.

4.3 Minimum Spanning Tree

Minimum spanning tree (MST) algorithms find an MST[1] of some arbitrary graph. Like the shortest-path problem, it has a polynomial-time, deterministic algorithm (i.e., it resides in **P**). The first MST algorithm [29] required

$$O(|E| \log |V|) \tag{4.5}$$

runtime. Numerous variations have since been proposed, with little change to the underlying scaling.

Because MST algorithms are efficient, they play a very useful role in the design of real-world network topologies, where resource minimisation is crucial.

4.4 Minimum-Cost Flow

The *minimum-cost flow problem* is that of minimising costs through a network for a specified amount of flow (i.e., total bandwidth or throughput), which acts as a constraint on the problem. The definition of 'cost' in this context is compatible with our earlier definition of cost metrics (Definition 1).

This problem can be efficiently solved using linear programming. Specifically, cost metrics along links in series are given by linear combinations of individual link costs. If, in addition, we let our net cost function be linear in the constituent costs, then the net cost will also be linear in all of the edge weights. This lends itself directly to optimisation via linear programming techniques. Algorithms for linear programming, such as the *simplex* algorithm, have polynomial-time solutions (i.e., reside in **P**), and a plethora of software libraries are available for implementing them numerically.

4.5 Maximum Flow

The *maximum flow problem* is the seemingly simple goal of – as the name suggests – maximising network flow, without consideration for any of the other cost metrics associated with the network. This type of problem is relevant when brute bandwidth is the dominant goal.

This problem can be tackled using a number of techniques. In some circumstances, linear programming techniques can be employed. The best-known algorithm is the Ford-Fulkerson algorithm, which finds a solution in

$$O(|E| \cdot c_{\max}) \tag{4.6}$$

---

[1] There may be multiple distinct MSTs for a given graph.

runtime, where $|E|$ is the number of links in the network and c_{max} is the maximum cost present in the network. The algorithm behaves pathologically in some conditions, which can easily be overcome in the context we present here. Using Ford-Fulkerson as a starting point, numerous other more sophisticated algorithms have been developed.

4.6 Multicommodity Flow

The *multicommodity flow problem* generalises the previous algorithms to be applicable to multi-user networks. The generalisation is that there may be a number of distinct senders residing on different nodes, each transmitting to distinct recipients residing on different nodes. This is the most realistic scenario we are likely to encounter in a real-world quantum internet, where networks will inevitably be shared by many users residing at different nodes.

Unfortunately, the computational complexity of solving this problem is much harder than the previous algorithms in general. Specifically, solving the problem exactly is **NP**-complete in general. However, in specific circumstances it can be approached using linear programming or polynomial-time approximation schemes.

4.7 Vehicle Routing Problem

The vehicle routing problem (VRP) is a multi-user generalisation of the shortest-path problem, where the goal is to minimise total network cost (i.e., the sum of all individual users' costs) when there are multiple users sharing the network, each with distinct sources and destinations (Figure 4.2).

Unlike the polynomial-time shortest-path algorithm, exactly solving the VRP is **NP**-hard in general. However, heuristic methods can find approximate suboptimal solutions far more efficiently, and there is a multitude of software packages available for doing so.

The VRP has found widespread use in, for example, the routing of transport networks for delivery companies or public transportation networks (hence the name), and many commercial companies exist that perform these kinds of optimisations on behalf of transport providers to enhance their efficiency.

It is obvious that this algorithm is directly applicable to multi-user communications networks, which are conceptually identical to transport networks, albeit a bit faster.

A multitude of variations on the VRP exist, accommodating different types of constraints (or additional flexibilities) in the operation of the network.

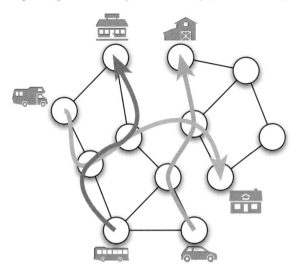

Figure 4.2 Example of the vehicle routing problem, the multi-user generalisation of the shortest-path problem, where the goal is to minimise the total cost across all users. Intuitively, because whenever individual shortest paths intersect one must trial different prioritisations, the combinatorics of this grow exponentially with the number of competing users.

4.8 Vehicle Rescheduling Problem

The vehicle rescheduling problem (VRSP) generalises the VRP to the case where properties of the network undergo changes dynamically within the course of trans- missions over the network. To use the analogy of transport networks, this could entail, for example, a truck breaking down en route to its destination, requiring real-time rescheduling of the other vehicles.

Solving the VRSP exactly is **NP**-complete in general but, as with the VRP, heuristic methods can often be applied that efficiently find approximate solutions.

In the context of communications over networks, the VRSP has obvious appli- cability – a quantum internet is likely going to be largely ad hoc in nature, with users coming and going and many nondeterministic points of failure, requiring ongoing updating of routing decisions if resource allocation is to remain as efficient as possible.

4.9 Improving Network Algorithms Using Quantum Computers

Given that we are directing this work at the upcoming quantum era, where quantum computing will become a reality, it is pertinent to ask whether quantum computers might improve the aforementioned network algorithms, some of which are computationally hard problems. Most notable, several of the discussed algorithms are **NP**-complete in general, a complexity class strongly believed to be exponentially complex on classical computers. Can quantum computers help us out here and improve network resource allocation? Can quantum computers help themselves?

Though it is not believed that quantum computers can efficiently solve such **NP**-complete problems, it is known that they can offer a quadratic speedup using Grover's unstructured search algorithm. Specifically, **NP**-complete problems can be treated as satisfiability problems, where we are searching for an input to a classical algorithm that yields a particular output.

To gain a quantum advantage, we treat the classical algorithm as an oracle whose input configurations form an unstructured search space. Then, Grover's algorithm can perform a search over the space of input configurations to find a satisfying solution, with quadratically enhanced runtime.

Though a quadratic improvement is far short of the exponential improvement we might hope for, Grover's algorithm is known to be optimal for the unstructured search problem. Nonetheless, despite only yielding a quadratic improvement, a quadratic speedup may already be sufficient to significantly improve network resource allocation.

Part II

Quantum Networks

Quantum networks comprise all of the same ingredients as classical networks but with some very important nonclassical additions. Nodes can additionally implement quantum computations, quantum-to-classical interfaces (i.e., measurements), quantum-to-quantum interfaces (i.e., switching data between different physical systems), quantum memories or any quantum process in general. Many of these are not allowed by the laws of classical physics.

The cost vectors associated with links could include measures that are uniquely quantum, such as fidelity, purity or entanglement measures, none of which are applicable to classical digital data.

As in the classical case, our goal is to find routing strategies that optimise a chosen cost measure. But in the quantum context costs will be constructed entirely differently owing to the quantum nature of the information being communicated.

We envisage a network with a set of senders and receivers, all residing on a time-dependent network graph as before. Senders have sets of quantum states they wish to communicate. For each state they must choose appropriate strategies, such that the overall cost is optimised, for some appropriate cost measure. Compared to classical resources, equivalent quantum resources are costly and must be used efficiently and frugally. Indeed, the no-cloning theorem imposes the constraint that arbitrary unknown states cannot be replicated at all! This makes resource allocation strategies of utmost importance in the quantum world.

Routing strategies will not always guarantee that packets have immediate access to network bandwidth the moment they demand it. One needs to think about the others, too! Inevitably, in shared networks there will sometimes be competition and congestion, forcing some users to wait their turn. For this reason, many quantum networks will require at least some nodes (the ones liable to competition) to have access to quantum memories, such that quantum packets can be buffered for a sufficient duration that they can wait their turn on the shared network resources for which there is high competition. The required lifetime of a quantum memory will then be related to overall network congestion. Of course, quantum memories induce unwanted quantum processes of their own, which need to be factored into cost calculations.

Given that classical networking is decades more advanced than quantum networking, and extremely cheap and reliable in comparison, we will assume that classical resources 'come for free' and only quantum resources are of practical interest in terms of their cost. That is, classical communication and computation is a free resource available to mediate the operation of the quantum network. We therefore envisage a *dual network* with two complementary networks operating in parallel and in tandem – the quantum network for communicating quantum data and a topologically identical classical network operating side by side and synchronised with the quantum network, overseeing and mediating the quantum network.

Data packets traversing the network will comprise both quantum and classical fields, which will be separated to utilise the appropriate network but synchronised such that they arrive at their destination as a single package of joint quantum and classical information to be at the disposal of the recipient.

The motivation for the dual network is to ensure that classical and quantum data that jointly represent packets remain synchronised and subject to the same QoS issues, such as packet collisions and network congestion.

We envisage quantum networks to extend beyond just client/server quantum computation, to include the free trade of any quantum asset. This includes state preparation, measurement, computation, randomness, entanglement and information. Much like the classical internet, by allowing quantum assets to be exchanged, we can maximise utility, improve economy of scale and enable new models for commercialisation.

May the games begin.

5

Quantum Channels

Like classical channels, quantum channels are inevitably subject to errors. These errors could be an intrinsic part of the system or induced by interaction with the external environment. The *quantum process* formalism provides an elegant mathematical description for all physically realistic error mechanisms [127, 74]. Here we review the quantum process formalism and how it applies to quantum networks. This paves the way for the quantum notion of costs.

5.1 Quantum Processes

To quantify the operation of nodes and links within our network, we must characterise the evolution they impose upon quantum states passing through them. Quantum processes, also known as *trace-preserving, completely positive maps* (CP-maps) are able to capture all of the physical processes relevant to quantum networking, such as unitary evolution, decoherence, measurement, quantum memory, state preparation, switching and, indeed, entire quantum computations. And they are able to capture physical processes in any degree of freedom, most commonly in the qubit basis, but also, for photons, in the spatiotemporal, photon-number, phase-space, or polarisation degrees of freedom.

Quantum processes are most easily represented using *Kraus operators*, $\{\hat{K}_i\}$,

$$\mathcal{E}(\hat{\rho}) = \sum_i \hat{K}_i \hat{\rho} \hat{K}_i^\dagger, \qquad (5.1)$$

where

$$\sum_i \hat{K}_i^\dagger \hat{K}_i = \hat{I} \qquad (5.2)$$

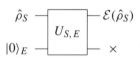

Figure 5.1 Model for the quantum process formalism, as a system state $\hat{\rho}_S$ undergoing joint unitary evolution with an environment state $|0\rangle_E$, which is subsequently traced out, yielding an arbitrary quantum process $\mathcal{E}(\hat{\rho}_S)$ acting on the primary system. For the most general possible class of quantum processes to be enabled, the dimension of the environment Hilbert space must grow quadratically with the dimension of the primary Hilbert space.

for normalisation. Here $\mathcal{E}$ is a superoperator, denoting the action of the process on state $\hat{\rho}$. This is also referred to as the *operator-sum representation*. This representation has the elegant interpretation as the probabilistic application of each of the Kraus operators $\hat{K}_i$, with probability

$$p_i = \mathrm{tr}(\hat{K}_i \hat{\rho} \hat{K}_i^\dagger). \tag{5.3}$$

In the ideal case, the two types of evolution of interest are unitary evolution, in which case there is only one Kraus operator, $\hat{K}_1 = \hat{U}$, and projective measurement, where there is again only one Kraus operator, $\hat{K}_1 = |m\rangle\langle m|$, for measurement outcome m.

Mathematically, quantum processes are equivalent to a state jointly undergoing unitary evolution with an external environment that is not observed (i.e., traced out),

$$\mathcal{E}(\hat{\rho}_S) = \mathrm{tr}_E(\hat{U}_{S,E}[\hat{\rho}_S \otimes |0\rangle_E\langle 0|_E]\hat{U}_{S,E}^\dagger), \tag{5.4}$$

where S denotes the primary system to which the process is applied, and E is an auxiliary environment system, as shown in Figure 5.1.

We will require that all of our states are normalised,

$$\mathrm{tr}(\hat{\rho}) = 1, \tag{5.5}$$

and that our processes are *trace preserving*. That is, they preserve normalisation

$$\mathrm{tr}[\mathcal{E}(\hat{\rho})] = 1. \tag{5.6}$$

Multiple consecutive processes may be composed using the notation

$$\mathcal{E}_n(\ldots \mathcal{E}_2(\mathcal{E}_1(\hat{\rho}))) = (\mathcal{E}_n \circ \cdots \circ \mathcal{E}_2 \circ \mathcal{E}_1)(\hat{\rho}). \tag{5.7}$$

In general, processes do not commute; i.e., $\mathcal{E}_1 \circ \mathcal{E}_2 \neq \mathcal{E}_2 \circ \mathcal{E}_1$. Unless unitary, quantum processes are irreversible, meaning that errors accumulate and cannot be overcome without the overhead of some form of quantum error correction (QEC)

[166, 40, 127]. The linearity of Eq. (5.1) implies that quantum processes are also linear,

$$\mathcal{E}(\hat{\rho}_1 + \hat{\rho}_2) = \mathcal{E}(\hat{\rho}_1) + \mathcal{E}(\hat{\rho}_2). \tag{5.8}$$

The only limitation faced by the quantum process formalism is that it is described over discrete time only. To consider continuous-time evolution, *master equations* can be used. These represent the continuous-time evolution of a quantum state as a differential equation in time, combining a usual Hamiltonian term as well as decoherence terms,

$$\frac{d\hat{\rho}}{dt} = -\frac{i}{\hbar}[\hat{H}, \hat{\rho}] + \sum_j (2\hat{L}_j \hat{\rho}\hat{L}_j^\dagger - \{\hat{L}_j^\dagger \hat{L}_j, \hat{\rho}\}), \tag{5.9}$$

where $\hat{H}$ is the Hamiltonian of the isolated system undergoing coherent evolution, and $\hat{L}_j$ are the *Lindblat operators*, capturing the incoherent component of the dynamics (i.e., environmental couplings). Here $[\cdot, \cdot]$ and $\{\cdot, \cdot\}$ are the commutator and anti-commutator, respectively.

In this work we will only make use of discrete-time quantum processes, because they naturally correspond to the evolution of states between discrete points within a network – we are typically only interested in the process undergone by a state from one end of a link to another, not the continuous-time dynamics of what takes place within them.

5.2 Quantum Process Matrices

In general, the Kraus operator representation for quantum processes is not unique – there may be multiple choices of Kraus operators that implement identical physical processes. But if the representation is not unique, how do we compare different quantum processes? To address this, it is common to choose a 'standard' basis for representing quantum processes, such that they may be consistently and fairly compared. This requires choosing a basis that is complete for operations on the Hilbert space acted upon by the process.

For example, for a single qubit, the Pauli operators – $\hat{\sigma}_1$ (identity, $\hat{I}$), $\hat{\sigma}_2$ (bit-flip, $\hat{X}$), $\hat{\sigma}_3$ (bit-phase-flip, $\hat{Y}$) and $\hat{\sigma}_4$ (phase-flip, $\hat{Z}$) – are complete for single-qubit operations (C_2). Therefore, by decomposing our Kraus operators into linear combinations of these basis operators we have a standardised representation for single-qubit processes. Formally, for one qubit,

$$\mathcal{E}(\hat{\rho}) = \sum_{i,j=1}^{4} \chi_{i,j} \hat{\sigma}_i \hat{\rho} \hat{\sigma}_j^\dagger. \tag{5.10}$$

The Hermitian matrix χ is known as the *process matrix*, from which many other metrics of interest may be directly computed (some of which are discussed in Chapter 8).

Process matrices share many algebraic properties and interpretations in common with density matrices. The diagonal elements can be regarded as the amplitudes associated with applying each of the four Pauli operators, all of which are nonnegative, whereas the off-diagonal elements represent the coherences between them; i.e., whether the operations on the diagonal are being applied probabilistically or coherently. For example, a process that simply randomly applies Pauli operators would have a diagonal process matrix in the Pauli basis. But off-diagonal elements would be indicative of applying coherent superpositions of the operators. Like density matrices, the dimensionality of process matrices grows exponentially with the number of qubits in the system being characterised and for exactly the same conceptual reasons.

For the process to be trace preserving we require

$$\mathrm{tr}(\chi) = 1. \tag{5.11}$$

We will typically enforce this constraint on our processes. $\mathrm{tr}(\chi) < 1$ implies non-determinism; i.e., the process sometimes fails.

As an illustrative example of the interpretation of process matrices, in Figure 5.2 we show the process matrix for the controlled-NOT (CNOT) gate, represented in the Pauli basis. The CNOT operator can be expressed in the Pauli operator basis as

$$\hat{U}_{\mathrm{CNOT}} = \frac{1}{2}(\hat{I} \otimes \hat{I} + \hat{I} \otimes \hat{X} + \hat{Z} \otimes \hat{I} - \hat{Z} \otimes \hat{X}). \tag{5.12}$$

Then, some density operator evolved under the CNOT gate is simply

$$\hat{U}_{\mathrm{CNOT}} \hat{\rho} \, \hat{U}_{\mathrm{CNOT}}^{\dagger}. \tag{5.13}$$

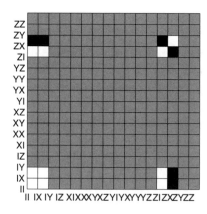

Figure 5.2 Process matrix for the CNOT gate, expressed in the Pauli basis. Colour coding: grey $= 0$, white $= 1/4$, black $= -1/4$.

Expanding this out, we obtain a new state comprising 16 terms, each representing the action of some combination of Pauli operators from the left and from the right. The amplitudes of these terms exactly correspond to the 16 nonzero elements of the process matrix shown in Figure 5.2.

5.3 Quantum Processes in Quantum Networks

Letting v_i represent the ith node within a route R, the process associated with communication from that node to the next is $\mathcal{E}_{v_i \to v_{i+1}}$. For the same network used previously, Figure 5.3 shows the quantum processes associated with the links in the network. The cumulative process associated with an entire route is therefore

$$\mathcal{E}_R = \mathcal{E}_{v_{|R|-1} \to v_{|R|}} \circ \cdots \circ \mathcal{E}_{v_2 \to v_3} \circ \mathcal{E}_{v_1 \to v_2}, \tag{5.14}$$

where $|R|$ is the number of nodes in R, and to simplify notation, all v_i are implicitly over the route R.

In general, both nodes and links in a quantum network may implement quantum processes. However, for the purposes of compatibility with the graph-theoretic algorithms described in Chapter 4, we will eliminate node processes by merging them into link processes, such that the processes in the network are described entirely by links. This reduction procedure is straightforward, shown in Figure 5.4.

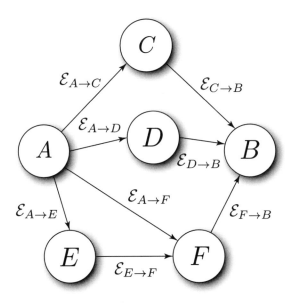

Figure 5.3 The network from Figure 2.1, with the quantum processes associated with each link. The net process associated with a route is given by the composition of each of the processes over the length of the route. For example, the route $R_1 = A \to C \to B$ induces the process $\mathcal{E}_{R_1} = \mathcal{E}_{C \to B} \circ \mathcal{E}_{A \to C}$.

(a)
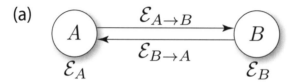

(b)

Figure 5.4 Removing node processes from network graphs on a trivial network with two nodes, A and B. Each node is associated with a quantum process ($\mathcal{E}_A$ and $\mathcal{E}_B$). Similarly, each link is associated with a process ($\mathcal{E}_{A \to B}$ and $\mathcal{E}_{B \to A}$). (a) Representation where the node and link processes are shown explicitly. (b) The node processes are replaced with identity operations by replacing each link process with the composition of the link process and its target node process. Equivalently, the cost of each node process is added to the cost of *every* incoming link and then eliminated. This procedure requires that all links be directed. If undirected links are present, they may simply be replaced by two directed links, one in each direction, implementing identical quantum processes each way.

5.4 Characterising Quantum States and Channels

Given a link implementing some arbitrary quantum process, it is essential that it can be experimentally determined such that network performance may be characterised. For example, if an optical channel is lossy, what is the loss rate? This is crucial when attempting to choose routing strategies that optimise certain cost metrics.

Treating a link or node as an unknown black box, *quantum process tomography* (QPT) [45] is a technique that may be applied to fully characterise the quantum process it implements, reproducing its complete process matrix. QPT has become a standard procedure, demonstrated in numerous architectures, most notably in optics [128, 149].

QPT works in general for processes in any degree of freedom; e.g., the qubit degree of freedom. However, it is important to note that full QPT requires statistics across the entire basis over which measurements are defined, which typically grows exponentially with the size of the system. For example, the number of measurement bases required to perform full QPT on n qubits grows exponentially with n.

However, often full process characterisation is not necessary. Instead, knowing particular metrics of interest may suffice. Some of the more noteworthy such metrics will be discussed in Chapter 8. In this instance, much work has been done in the field of *compressed sensing* or *compressed quantum process tomography*, in which some process metrics of interest can be experimentally determined using far fewer physical resources (with efficient scaling!) than via a full reconstruction of the process matrix using QPT. As a most trivial example, if the loss associated with a fibre-optic channel is the metric of interest, this can be much more easily determined than by performing full QPT.

On the other hand, however, most quantum channels are designed to accommodate systems with very limited Hilbert space dimensionality per clock-cycle – e.g., a fibre-optic link might transmit just one photon at a time – in which case there is no exponentiality to be terribly concerned about (QPT of a single-photon channel is trivial).

Importantly, it is often the case that the quantum process associated with a channel will remain constant over time. The efficiency of a length of fibre, for example, does not change. In this instance, characterising the channel need only be performed once in advance, without requiring ongoing dynamic updating. On the other hand, when communicating with satellites in low Earth orbit it is to be expected that the properties of links will be highly dynamic.

We will now explain QPT in the archetypal context of single-qubit channels, which logically generalises to multiple qubits and can similarly be generalised to non-qubit systems also.

Quantum State Tomography

The first stage in QPT is *quantum state tomography* (QST), where the goal is to reconstruct an unknown density matrix via measurements upon multiple copies of the state. QST is based on the simple observation that the completeness relation for an arbitrary state can be expressed,

$$\hat{\rho} = \sum_i \text{tr}(\hat{E}_i \hat{\rho}) \cdot \hat{E}_i, \tag{5.15}$$

where $\{\hat{E}_i\}$ forms a complete basis for the Hilbert space of $\hat{\rho}$. For a single qubit this decomposition is most often performed in the Pauli basis,

$$\hat{\rho} = \text{tr}(\hat{\rho}) \cdot \hat{I} + \text{tr}(\hat{X}\hat{\rho}) \cdot \hat{X} + \text{tr}(\hat{Y}\hat{\rho}) \cdot \hat{Y} + \text{tr}(\hat{Z}\hat{\rho}) \cdot \hat{Z}$$

$$= \sum_{i=1}^{4} \text{tr}(\hat{\sigma}_i \hat{\rho}) \cdot \hat{\sigma}_i, \tag{5.16}$$

where σ_i denote the four Pauli operators. Of course, $\text{tr}(\hat{E}\hat{\rho}) = P(\hat{E}|\hat{\rho})$ is just the expectation value of the measurement operator $\hat{E}$ when measuring $\hat{\rho}$. Thus, measuring the expectation values in each of the four Pauli bases reconstructs $\hat{\rho}$.

This generalises straightforwardly to multi-qubit systems, where we measure all combinations of tensor products of the Pauli operators, the number of which grows exponentially with the number of qubits n, as 4^n. This introduces scalability issues for systems comprising a large number of qubits.

In the case of optical systems, entirely alternate, but equivalent, approaches may be used, such a probing the Wigner function directly using homodyne detection.

Quantum Process Tomography

Now to perform QPT we apply the unknown process to a complete basis of input states $\{\hat{\rho}_i\}$ and perform QST on the output state for each. This yields

$$\mathcal{E}(\hat{\rho}_j) = \sum_i c_{i,j} \hat{\rho}_i, \tag{5.17}$$

where the sum runs over the basis of states. From QST, all of the coefficients $c_{i,j}$ may be determined. Next we define the following decomposition for each of the terms in the sum of Eq. (5.10),

$$\hat{E}_m \hat{\rho}_j \hat{E}_n^\dagger = \sum_k B_{j,k}^{m,n} \hat{\rho}_k, \tag{5.18}$$

where B defines a decomposition in the chosen basis, not dependent on any measurement results. Then we can write

$$\mathcal{E}(\hat{\rho}_j) = \sum_{m,n} \chi_{m,n} \hat{E}_m \hat{\rho}_j \hat{E}_n^\dagger$$

$$= \sum_{m,n} \sum_k \chi_{m,n} B_{j,k}^{m,n} \hat{\rho}_k. \tag{5.19}$$

Because $\hat{\rho}_k$ form a linearly independent basis, we can write the decomposition

$$c_{j,k} = \sum_{m,n} \chi_{m,n} B_{j,k}^{m,n}, \tag{5.20}$$

for all j,k. From this, standard linear algebra techniques allow an inversion to obtain

$$\chi_{m,n} = \sum_{j,k} (B_{j,k}^{m,n})^{-1} c_{j,k}, \tag{5.21}$$

thereby obtaining the full process matrix χ in the chosen basis.

6

Optical Encoding of Quantum Information

Though all-optical quantum computing is an unlikely architecture for future scalable quantum computers, it is all but inevitable that optics will play a central role in quantum communications networks. Foremost, this is because photons are 'flying' by their very nature and can very easily be transmitted across large distances – it is quite challenging to transmit a superconducting circuit containing information from Australia to Mozambique in the blink of an eye! Additionally, optical states are, in many cases, relatively easy to prepare, manipulate and measure and can also be readily interfaced with other physical quantum systems, allowing the transfer of quantum information from optical communications systems to some other architecture better suited to a given task.

Optical systems are very versatile, allowing quantum information to be optically encoded in a number of ways – into single photons, many photons or even an indeterminate number of photons, and in both discrete or continuous degrees of freedom. Different types of encodings may have very different properties in terms of the errors they are susceptible to.

When dealing with single photons, information can be encoded in a number of ways. Most obviously, it can be encoded into the polarisation basis, allowing one qubit of information per photon (i.e., horizontal and vertical polarisation represent the logical $|0\rangle$ and $|1\rangle$ states). Or it could be directly encoded into the photon-number basis. However, other degrees of freedom, such as the spectral/temporal degrees of freedom could be employed, encoding information into time or frequency bins, with potentially far more levels than a simple polarisation qubit [146]. Next we discuss the main methods for optical encoding of quantum information.

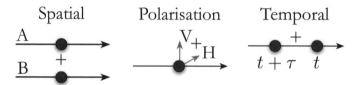

Figure 6.1 Three approaches to encoding a single qubit using a single photon, via a superposition across two spatial (A and B), polarisation (V and H) or temporal (t and $t + \tau$) modes.

6.1 Single Photons

A very attractive feature of single photons is that they undergo very little decoherence, even over large distances – dephasing in the polarisation degree of freedom, for example, is negligible in free-space. They are, however, very susceptible to loss, and protocols relying on many single-photon states suffer exponential decay in their success rates as the number of photons is increased.

We can encode a single qubit into a single photon in the polarisation basis using the horizontal and vertical polarisation degrees of freedom. Equivalently, one can employ 'dual rail' encoding, whereby a single photon is placed into a superposition across two spatial modes. Finally, one can use time-bin encoding, whereby discrete windows of time represent logical basis states when occupied by a photon. This leads to the equivalent representations for logical qubits (L),

$$|\psi\rangle_{\text{qubit}} \equiv \alpha|0\rangle_L + \beta|1\rangle_L,$$

$$|\psi\rangle_{\text{pol}} \equiv \alpha|H\rangle + \beta|V\rangle,$$

$$|\psi\rangle_{\text{dual}} \equiv \alpha|0,1\rangle + \beta|1,0\rangle,$$

$$|\psi\rangle_{\text{temporal}} \equiv \alpha|0_t, 1_{t+\tau}\rangle + \beta|1_t, 0_{t+\tau}\rangle, \tag{6.1}$$

shown graphically in Figure 6.1.

Conversion between polarisation and dual-rail encoding is straightforward and deterministic using standard optical components, as described in Figure 6.2.

Note that polarisation encoding requires a single spatial mode per qubit, whereas dual-rail encoding requires two. Polarisation encoding brings with it the advantage that arbitrary single-qubit operations may be implemented using wave plates, which maintain coherence between the basis states extraordinarily well. In dual-rail encoding, on the other hand, single-qubit operations are implemented using beamsplitter operations between the two spatial modes, which must be interferometrically stable, because consecutive single-qubit operations yield Mach-Zehnder (MZ) interference [192, 193], to be discussed in detail in Section 12.2.

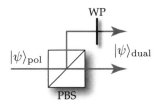

Figure 6.2 Conversion from single-photon polarisation encoding to dual-rail encoding, using a polarising beamsplitter (PBS) and wave plate (WP). The PBS separates the polarisation components into two distinct spatial modes. The WP then rotates the polarisation of one of the spatial modes such that it has the same polarisation as the other. Conversion from dual-rail to polarisation encoding is just the reverse of this procedure.

Single-photon encodings are extremely important, because they form the basis for universal linear optics quantum computing, BOSONSAMPLING and quantum walks. They are also the simplest optical states for representing single qubits.

6.2 Photon Number

Of course, the photon-number degree of freedom need not be limited to 0 or 1 photons. By fully exploiting the photon-number degree of freedom, we can encode a qudit[1] of arbitrary dimension into a single optical mode,

$$|\psi\rangle_{\text{qudit}} \equiv \sum_{n=0}^{\infty} \alpha_n |n\rangle. \tag{6.2}$$

This may give the impression that a single optical mode has infinite information capacity. Needless to say, this sounds too good to be true, and it is. Loss decoheres photon-number-encoded states exponentially with photon number, because for large photon number the probability of a number state retaining its photon number exponentially asymptotes to zero. So although in principle we can encode an ∞-level qudit, the moment any nonzero loss is introduced, this exponential dependence destroys the state.

Though photon-number encoding can be useful for communications purposes, it is not very practical for quantum information processing tasks, because operations between basis states are not energy preserving, with each basis state having energy $E = n\hbar\omega$, where ω is frequency and $\hbar$ is Planck's constant. Thus, qudit operations would need to be active processes.

---

[1] A d-level system, as opposed to a qubit's two levels.

6.3 Spatiotemporal

Completely independent of the photon-number degree of freedom are the spatiotemporal degrees of freedom, which encode the spatial, temporal or spectral structure of photons. In the temporal domain, for example, we could define the temporal structure of a single photon as

$$|\psi\rangle_{\text{temporal}} = \int_{-\infty}^{\infty} \psi(t)\hat{a}^\dagger(t)\,dt\,|0\rangle, \tag{6.3}$$

where $\hat{a}^\dagger(t)$ is the time-specific photonic creation operator, and $\psi(t)$ is the temporal distribution function [150]. Equivalently, one could take the Fourier transform of the temporal distribution function and represent the same state in the frequency basis

$$\tilde{\psi}(\omega) = \mathcal{FT}(\psi(t)). \tag{6.4}$$

Likewise, one could employ a similar representation in the transverse spatial degrees of freedom, with spatial distribution function $\psi(x, y)$.

Alternatively, we can define *mode operators* [148], which are mathematically equivalent to creation operators but create photons with a specific temporal envelope,

$$\hat{A}^\dagger_\psi = \int_{-\infty}^{\infty} \psi(t)\hat{a}^\dagger(t)\,dt,$$

$$|\psi\rangle_{\text{temporal}} = \hat{A}^\dagger_\psi|0\rangle. \tag{6.5}$$

Mode operators commute, inheriting this property directly from photonic creation operators,

$$\left[\hat{A}^\dagger_{\psi_1}, \hat{A}^\dagger_{\psi_2}\right] = 0. \tag{6.6}$$

Now by defining an orthonormal basis of temporal distribution functions, $\{\xi_i\}$, such that

$$\langle 0|\hat{A}_{\xi_i}\hat{A}^\dagger_{\xi_j}|0\rangle = \delta_{i,j}, \tag{6.7}$$

we can encode a qudit of arbitrary dimension into the spatiotemporal degrees of freedom,

$$|\psi\rangle_{\text{qudit}} \equiv \sum_{i=0}^{\infty} \alpha_i \hat{A}^\dagger_{\xi_i}|0\rangle. \tag{6.8}$$

This encoding allows a qudit of arbitrary dimension to be encoded into a single spatial mode. Again, however, summing to infinity is somewhat fanciful, given any physically realistic spatiotemporal error model, such as an imperfect frequency response in the channel; e.g., a bandpass response of an optical fibre or photodetector.

The spectral basis functions could take the form of any orthonormal basis of complex functions, such as wavelets, Hermite functions or well-separated functions with finite support.

Time Bins

In time-bin encoding we define our basis of modes (whether qubits or higher-dimensional qudits) as distinct, nonoverlapping time bins, which are localised wave packets in the temporal degree of freedom, each separated from the next by some fixed interval τ. This can be considered a special case of spatiotemporal encoding, where the basis mode functions satisfy the relation

$$\xi_j(t) = \xi_0(t - j\tau), \tag{6.9}$$

as well as the usual orthonormality constraints. Here τ is sufficiently large, and $\xi_i(t)$ sufficiently temporally localised, that the temporal modes are orthogonal as per Eq. (6.7). The encoding is shown graphically in Figure 6.3.

Time-bin encoding arises naturally in architectures where the photon source driving the system is operating at a high repetition rate, R, in which case $\tau = 1/R$. Architectures for optical quantum computing have been described [115, 144] and experimentally demonstrated based entirely on time-bin encoding.

These schemes can be very resource efficient, because a single source operating at a high repetition rate can replace an entire bank of distinct sources that would ordinarily be required in spatial architectures. Similarly, a single time-resolved detector, with resolution at least τ, can replace a bank of detectors operating in parallel. And only a single spatial mode is required to store an arbitrary number of qubits/qudits, as long as it is long enough to support the entire pulse-train – at least $2n\tau$ for n qubits.

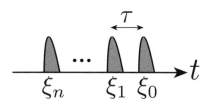

Figure 6.3 Time-bin encoding on an n-level quantum system. Time bins are temporally localised with envelopes ξ_j, orthogonal to all others, different only via temporal displacements. The temporal separation between each consecutive time bin is given by τ.

In the schemes of [115, 144], entire optical quantum computing protocols can be efficiently constructed using only a single source, a single detector, two delay-lines and three dynamically controlled beamsplitters, irrespective of the size of the computation, an enormous resource savings compared to traditional spatial encodings. Furthermore, in these schemes, there is only a single point of interference, greatly simplifying optical interferometric alignment, which would ordinarily require simultaneously aligning a large number of optical elements, as many as $O(m^2)$ elements for an m-mode network [139].

6.4 Phase Space

When encoding information optically, we need not restrict ourselves to photon-number states (discrete variables). We also have a lot of flexibility to encode information in phase space using continuous-variable (CV) states, where phase and amplitude relations encode quantum information [39]. In this formalism, rather than expressing states in terms of photonic creation operators, $\hat{a}^\dagger$, we represent them using phase-space position ($\hat{r}$) and momentum ($\hat{p}$) operators.

In phase space, the most common method for visualising optical states is in terms of quasi-probability functions,[2] of which there are a multitude. The best known quasi-probability representations are the following:

- P-function: represents a state as a quasi-mixture of coherent states. When the P-function is strictly nonnegative, it can be interpreted as a perfect classical mixture of coherent states. However, with any negativity this classical interpretation breaks down; hence 'quasi'-probability. In general, the P-function representation for a state is not unique.

$$\hat{\rho} = \iint P(\alpha)|\alpha\rangle\langle\alpha| d^2\alpha. \tag{6.10}$$

- Q-function: represents a state in terms of its overlap with the complete set of all coherent states, which form an overcomplete basis.

$$Q(\alpha) = \frac{1}{\pi}\langle\alpha|\hat{\rho}|\alpha\rangle. \tag{6.11}$$

---

[2] The term 'quasi-probability' arises because in some regimes (for example, strictly nonnegative P-functions), the function has a true probabilistic interpretation. However, this interpretation breaks down for any negativity in the $P(\alpha)$, because negative probabilities have no meaningful classical interpretation.

- Wigner function: also has a quasi-probability interpretation, and negativity is qualitatively associated with 'quantumness'. The Wigner function of a state is unique and isomorphic to the density operator, making it perhaps the most useful phase-space representation for quantum states of light.

$$W(x, p) = \int e^{ips/\hbar} \left\langle x - \frac{s}{2} \middle| \hat{\rho} \middle| x + \frac{s}{2} \right\rangle ds. \qquad (6.12)$$

These representations, though entirely equivalent to a photon-number basis representation, are far easier to work with for many types of states. Most notable, Gaussian states are conveniently represented and manipulated using phase-space representations.

Coherent states

As the most trivial CV encoding of quantum information, consider coherent states. These are particularly useful because they are pure states, with well-defined coherence relationships, and are closely approximated by laser sources and therefore readily available in the lab.

A coherent state, $|\alpha\rangle$, is parametrised by a single complex parameter, α, given by a phase and amplitude,

$$|\alpha\rangle = e^{-\frac{|\alpha|^2}{2}} \sum_{n=0}^{\infty} \frac{\alpha^n}{\sqrt{n!}} |n\rangle. \qquad (6.13)$$

Figure 6.4 illustrates the phase-space representation for two approximately orthogonal coherent state basis states.

By manipulating these parameters, information can be encoded into coherent states. We could, for example, define two coherent states of opposite phase to represent qubit basis states,

$$|0\rangle \equiv |\alpha\rangle,$$
$$|1\rangle \equiv |-\alpha\rangle. \qquad (6.14)$$

Note, however, that this representation for qubits is only approximate, because coherent states are not orthogonal,

$$\langle \alpha | \beta \rangle = e^{-\frac{1}{2}(|\alpha|^2 + |\beta|^2 - 2\alpha^*\beta)}$$
$$\neq \delta(\alpha - \beta); \qquad (6.15)$$

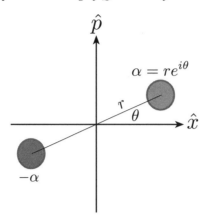

Figure 6.4 Phase-space representation of a coherent state with complex amplitudes $\pm\alpha$. For sufficiently large amplitude one can approximate orthogonality,
enabling a qubit encoding.

thus, the two logical basis states are not perfectly orthogonal,

$$\langle -\alpha | \alpha \rangle = e^{-2|\alpha|^2}, \tag{6.16}$$

which is nonzero for any finite α, whereas for ideal qubits we require $\langle 0|1 \rangle = 0$.
However, for large α, $| \pm \alpha \rangle$ closely approximate orthogonality, allowing them to
be used as qubits.

This representation for qubits using coherent states is easily generalised to qudits
by considering coherent states orbiting the origin of phase space at equal angular
intervals of $2\pi/d$, for a d-level qudit. The kth qudit basis state is then

$$|k\rangle_d = |\alpha e^{ik/d}\rangle, \tag{6.17}$$

for $k = 0, \ldots, d - 1$, where again the basis states are nonorthogonal but closely
approximate orthogonality for large α. The qudit value k can easily be manipulated
using simple phase-shift operators,

$$\hat{\Phi}(\phi) = e^{i\phi\hat{n}}, \tag{6.18}$$

where $\hat{n} = \hat{a}^\dagger\hat{a}$ is the photon-number operator. These phases are trivially implemented in the laboratory as wavelength-scale modulations in optical path length
(i.e., a thin piece of glass or other transmissive material with a different refractive
index).

Note that despite being pure states, with well-defined coherence, coherent states
are considered classical, because they are unable to encode quantum information.
That is, the coherence relationships cannot be exploited for the encoding of qubits
or qudits.

Coherent states are useful in that they are easy to prepare using modern lasers, including laser diodes, and by turning up the amplitude can be transmitted over long distances, with loss not affecting quantum coherence, only the amplitude.

Coherent state encoding can be regarded as encoding via the displacement operator

$$\hat{D}(\alpha) = \exp\left[\alpha\hat{a}^\dagger - \alpha^*\hat{a}\right],$$ (6.19)

which implements translations in phase space via the addition of coherent amplitude to a state. Coherent states are simply obtained as displaced vacuum states,

$$\hat{D}(\alpha)|0\rangle = |\alpha\rangle.$$ (6.20)

Squeezed States

In the same way that information can be encoded using the displacement operator via coherent states, we can encode information via the single-mode squeezing operator,

$$\hat{S}(\xi) = \exp\left[\frac{1}{2}(\xi^*\hat{a}^2 - \xi\hat{a}^{\dagger 2})\right],$$ (6.21)

where $\xi = re^{i\varphi} \in C$, r is known as the squeezing parameter, which will determine the magnitude of the squeezing, and $\varphi \in [0, 2\pi]$ denotes the axis along which the squeezing is taking place.

Graphically, in terms of their phase-space visualisation, squeezing implements dilations about a given axis. Strongly squeezing the vacuum state along the $\hat{x}$ or $\hat{p}$ directions yields two states that are approximately orthogonal for large squeezing amplitudes. Thus, with sufficient squeezing they can be used as a basis for *approximating* a qubit. This encoding can be exploited for full universal quantum computing.

6.5 Nonoptical Encoding

In a nonoptical context, the elementary unit of quantum information – the qubit – can be naturally encoded into any system with a natural or engineered two-level structure. This actually encompasses a broad range of possibilities, including, among many others:

- Two-level atoms: let two distinct electron energy levels, with long lifetimes, represent the two logical basis states.
- λ-configuration atoms: atoms with two degenerate ground states, which encode the logical qubit, and an additional excited state, which may be transitioned to

upon excitation from only one of the ground states. Relaxation from the excited state enables optical coupling via the emitted photon.

- Quantum dots: essentially artificial atoms, which can be engineered with custom band structures, allowing two- or higher-level qudits to be easily fabricated.
- Nitrogen-vacancy (NV) centres: a type of point defect in diamond, which has a very well-defined energy-level structure that may be utilised to represent qubits.
- Atomic ensembles: encode quantum information similarly to a single atom, except that the excitation is a *collective* one, in superposition across all atoms in the ensemble.
- Superconducting rings: a superposition of current flow direction in a superconducting ring represents the two logical basis states.
- Trapped ions: qubits are encoded into stable electronic states of electromagnetically trapped ions.

Clearly the nonoptical elements in a quantum network must somehow interface with optical states, such that communication is facilitated, discussed in Section 10.1.

7

Errors in Quantum Networks

As with classical data, quantum data are susceptible to corruption during transmission. However, in addition to all of the usual classical error models, quantum information is subject to further uniquely quantum errors. These errors can be represented using the quantum process formalism and fully characterised using QPT. We now briefly discuss several of the dominant errors arising in quantum systems, paying special attention to error models acting on qubits and optical states, because these are the most relevant in a quantum networking context.

7.1 Loss

Given that quantum communication links will typically be optical, the dominant error mechanism is likely to be loss. We let the *efficiency*, η, of an optical quantum process be the probability that a given photon entering the channel leaves the channel in the desired mode, or probability $1 - \eta$ of being lost. In the case of information encoded into single-photon states – e.g., using the polarisation degree of freedom – η corresponds exactly to the success probability of the communication.

When implementing protocols employing post-selection upon detecting all photons, the protocol will be nondeterministic, where loss dictates the protocol's success probability. Specifically, with n photons, each with efficiency η, the net post-selection success probability of the entire device is

$$P = \eta^n. \tag{7.1}$$

This implies that an exponential number of trials,

$$N = \frac{1}{P} = \frac{1}{\eta^n}, \tag{7.2}$$

is required in post-selected protocols. Clearly this exponential scaling is of concern, requiring demanding efficiencies in future large-scale implementations.

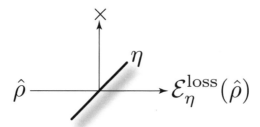

Figure 7.1 Model for the loss channel. The input state, $\hat{\rho}$, passes through a beamsplitter of transmissivity η, and the reflected mode is discarded, yielding the lossy output state $\mathcal{E}_\eta^{\text{loss}}(\hat{\rho})$.

Model

Formally, let $\mathcal{E}_\eta^{\text{loss}}$ be the loss channel with efficiency η. The channel acting on an initially pure single-photon state, $|1\rangle$, can be modelled as a beamsplitter with transmissivity η acting on the state, where the reflected mode is traced out, shown in Figure 7.1. This yields the quantum process

$$\mathcal{E}_\eta^{\text{loss}}(\hat{\rho}) = \text{tr}_B[\hat{U}_{\text{BS}}(\hat{\rho}_A \otimes |0\rangle_B\langle 0|_B)\hat{U}_{\text{BS}}^\dagger], \tag{7.3}$$

where $\hat{U}_{\text{BS}}$ is the beamsplitter operation.

Consecutive loss channels act multiplicatively (in the net efficiency) and commutatively,

$$\mathcal{E}_{\eta_1}^{\text{loss}} \circ \mathcal{E}_{\eta_2}^{\text{loss}} = \mathcal{E}_{\eta_2}^{\text{loss}} \circ \mathcal{E}_{\eta_1}^{\text{loss}} = \mathcal{E}_{\eta_1\eta_2}^{\text{loss}}. \tag{7.4}$$

Linear Optics Networks

In the special case of linear optics circuits, loss channels have the elegant property that, provided that the loss rate is uniform across all modes, they can be commuted through the circuit to the front or back. Specifically,

$$(\mathcal{E}_\eta^{\text{loss}})^{\otimes m} \circ \mathcal{E}_U = \mathcal{E}_U \circ (\mathcal{E}_\eta^{\text{loss}})^{\otimes m}, \tag{7.5}$$

where $\mathcal{E}_U$ is a unitary linear optics process, implementing a photon-number-preserving map of the form of Eq. (15.1). This is represented by the circuit diagram shown in Figure 7.2. This simplifies the treatment of distinct system inefficiencies (such as source, network and detector inefficiencies) by allowing us to commute them to the beginning or end of the circuit and combine them together into a single net efficiency. In many scenarios, this allows the different system inefficiencies to be dealt with via post-selection.

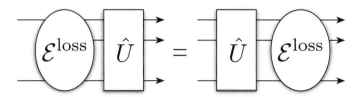

Figure 7.2 Commutation of a uniform loss channel (i.e., identical efficiency on each mode), $\mathcal{E}^{\text{loss}}$, through a passive linear optics network, $\hat{U}$.

Single-Photon Encoding

In the case of the vacuum and single-photon states, which is most common to qubit encodings, we obtain

$$\mathcal{E}_\eta^{\text{loss}}(|0\rangle\langle 0|) = |0\rangle\langle 0|,$$
$$\mathcal{E}_\eta^{\text{loss}}(|1\rangle\langle 1|) = (1-\eta)|0\rangle\langle 0| + \eta|1\rangle\langle 1|. \qquad (7.6)$$

This dynamic is of the same form as amplitude damping.

Polarisation and Dual-Rail Encoding

This process would apply equivalently to both horizontal and vertical polarisations. Therefore, via linearity, the loss channel acting on a polarisation-encoded qubit yields

$$\mathcal{E}_\eta^{\text{loss}}(|\psi\rangle_{\text{pol}}\langle\psi|_{\text{pol}}) = (1-\eta)|0\rangle\langle 0| + \eta|\psi\rangle_{\text{pol}}\langle\psi|_{\text{pol}}. \qquad (7.7)$$

The same applies in the context of dual-rail encoding. Note that though this transformation mixes the state in the photon-number degree of freedom, it preserves coherence between the horizontal and vertical single-photon components. Thus, upon successful post-selection, the state is projected back onto the desired qubit state.

Photon-Number Encoding

In the general case of an n-photon Fock state, we obtain

$$\mathcal{E}_\eta^{\text{loss}}(|n\rangle\langle n|) = \sum_{i=0}^{n} \binom{n}{i} \eta^i (1-\eta)^{n-i} |i\rangle\langle i|. \qquad (7.8)$$

In the case of higher order photon-number encoding of qudits, as per Eq. (6.2), the probability of an n-photon basis state being maintained scales as η^n. That is, if the highest photon-number term in our qudit is n, that component has an exponentially low probability of being preserved through the loss channel. For this fundamental reason, photon-number encoding does not enable infinite-dimensional qudits to be encoded.

Coherent State Encoding

Coherent states are the one example of states that are in a sense robust against loss, because a lossy coherent state is another coherent state with lower amplitude but without any loss in coherence,

$$\mathcal{E}_\eta^{\text{loss}}(|\alpha\rangle\langle\alpha|) = |\eta\alpha\rangle\langle\eta\alpha|. \tag{7.9}$$

This arises because coherent states are eigenstates of the photonic annihilation operator, $\hat{a}|\alpha\rangle = \alpha|\alpha\rangle$.

However, although coherence is maintained under the loss channel, the process is irreversible, because noise-free amplitude amplification is not possible in general. Thermal states exhibit the same property, that a loss channel simply yields another thermal state with reduced amplitude, although these exhibit no coherence.

Scaling

The scaling of loss over distance d varies depending on the medium through which the light traverses. We will consider two dominant mediums most relevant to future quantum networking:

- Optical fibre: mode geometry is well preserved, but optical medium is intrinsically lossy.
- Free-space: mode geometry is subject to dispersion but the medium is either lossless (in vacuum) or very low-loss (in atmosphere).

When propagating through fibre (or atmosphere, or some other lossy medium) net efficiency scales inverse exponentially as

$$\eta = O(e^{-\alpha d}), \tag{7.10}$$

where α is a characteristic of the medium.[1]

In free-space, on the other hand, where the medium of propagation is vacuum, which is effectively lossless, the effective loss rate is not determined by the medium but rather by the fact that the spot size of an optical state is subject to dispersion and grows only quadratically with distance, as shown in Figure 7.3. Then when the

---

[1] With present-day fibre technology, this characteristic decay rate is on the order of $\alpha = \frac{1}{22\text{km}}$.

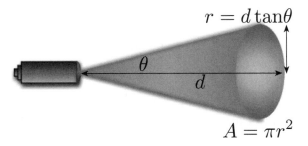

Figure 7.3 Spot size of an optical beam grows quadratically with distance from the source. If the beam is then collected via a camera with aperture area A', any component of A falling outside of A' is effectively lost, yielding an effective loss channel with efficiency $\eta = \frac{A'}{A} = \frac{A'}{\pi d^2 \tan^2 \theta} = O\left(\frac{1}{d^2}\right)$.

light is detected, if the spot size is greater than the detector aperture, the undetected component effectively translates to loss. Thus, through free-space the effective efficiency scales as

$$\eta = O\left(\frac{1}{d^2}\right),\tag{7.11}$$

which is far more favourable than the exponential scaling inherent to lossy mediums. This provides space-based quantum networks with an inherent competitive advantage compared to any form of ground-based network.

Through atmospheric channels we will have both dispersion and distance-dependent loss, yielding an effective loss rate given by the product of the two effects,

$$\eta = O\left(\frac{e^{-\alpha d}}{d^2}\right).\tag{7.12}$$

7.2 Dephasing

The dephasing error model describes the deterioration of quantum coherence in a state. It does not change the actual amplitudes of the components in the superposition but rather reduces the state to a mixture of those components. Thus, dephasing can be thought of as destroying quantum information (coherence) while retaining classical information (probability amplitudes).

Qubits

In terms of qubits, dephasing is most commonly represented using the Kraus representation,

$$\mathcal{E}_p^{\text{dephasing}}(\hat{\rho}) = p \cdot \hat{\rho} + (1-p) \cdot \hat{Z}\hat{\rho}\hat{Z}, \tag{7.13}$$

where $\hat{\rho}$ is the state of a single qubit, and $\hat{Z}$ is the Pauli phase-flip operator.[2] Intuitively this tells us that the dephasing channel creates a mixture of an input state with its phase-flipped self.

An alternate interpretation for the dephasing channel is that it is equivalent to the outside environment measuring $\hat{\rho}$ in the logical ($\hat{Z}$) basis but unknown to us, thereby projecting the state onto one basis state or another, yielding a mixture of the two.

Dephasing acting on $\hat{\rho}$ can be very elegantly visualised as simply nullifying the off-diagonal matrix elements; i.e., eliminating coherence terms. Dephasing is a ubiquitous error mechanism and affects all current quantum computing architectures.

Consecutive dephasing channels accumulate into another dephasing channel,

$$\mathcal{E}_{p_1}^{\text{dephasing}} \circ \mathcal{E}_{p_2}^{\text{dephasing}} = \mathcal{E}_{p'}^{\text{dephasing}}, \tag{7.14}$$

where

$$p' = p_1 p_2 + (1-p_1)(1-p_2); \tag{7.15}$$

i.e., the probability that an even number of phase-flips have occurred.

As a simple example, consider the $p = 1/2$ dephasing channel acting on the $|+\rangle = \frac{1}{\sqrt{2}}(|0\rangle + |1\rangle)$ state. Then we have

$$\mathcal{E}_{1/2}^{\text{dephasing}}(|+\rangle\langle+|) = \frac{1}{2}(|+\rangle\langle+| + \hat{Z}|+\rangle\langle+|\hat{Z})$$

$$= \frac{1}{2}(|+\rangle\langle+| + |-\rangle\langle-|)$$

$$= \frac{1}{2}(|0\rangle\langle0| + |1\rangle\langle1|)$$

$$= \frac{\hat{I}}{2} \tag{7.16}$$

is the completely mixed state. That is, the state has completely decohered. Note, however, that this complete decoherence depended on the choice of input state. A computational basis state, on the other hand, would be left unchanged by this channel,

---

[2] Bit-flip and bit-phase-flip channels may be represented similarly by replacing $\hat{Z}$ with $\hat{X}$ or $\hat{Y}$, respectively, although these do not arise as naturally as dephasing in many physical contexts.

$$\mathcal{E}_{1/2}^{\text{dephasing}}(|0\rangle\langle 0|) = \frac{1}{2}(|0\rangle\langle 0| + \hat{Z}|0\rangle\langle 0|\hat{Z})$$

$$= |0\rangle\langle 0|,$$

$$\mathcal{E}_{1/2}^{\text{dephasing}}(|1\rangle\langle 1|) = \frac{1}{2}(|1\rangle\langle 1| + \hat{Z}|1\rangle\langle 1|\hat{Z})$$

$$= |1\rangle\langle 1|. \tag{7.17}$$

Note that the probability of no dephasing occurring over multiple dephasing channels in series is given by the product of the respective probabilities for the individual channels.

T_2-Times

A qubit dephasing channel is often quoted in terms of its T_2-time, a characteristic time for dephasing to occur under continuous-time evolution. Specifically, the probability of no dephasing occurring scales as

$$p_{\text{no error}} = e^{-t/T_2}, \tag{7.18}$$

yielding the equivalent dephasing channel

$$\mathcal{E}_t^{\text{dephasing}}(\hat{\rho}) = e^{-t/T_2}\hat{\rho} + \frac{1}{2}(1 - e^{-t/T_2})(\hat{\rho} + \hat{Z}\hat{\rho}\hat{Z}). \tag{7.19}$$

For a qubit density matrix

$$\hat{\rho} = \begin{pmatrix} \alpha & \gamma \\ \gamma^* & \beta \end{pmatrix}, \tag{7.20}$$

this is equivalent to adding a factor of e^{-t/T_2} to the two off-diagonal (coherence) elements

$$\hat{\rho}_t = \begin{pmatrix} \alpha & e^{-t/T_2}\gamma \\ e^{-t/T_2}\gamma^* & \beta \end{pmatrix}. \tag{7.21}$$

Note that Eq. (7.19) is parametrised into an ideal term ($\hat{\rho}$) and a completely dephased term, $(\hat{\rho} + \hat{Z}\hat{\rho}\hat{Z})/2$, and thus multiple rounds of this channel yields an equivalent channel where the probability associated with the former term accumulates multiplicatively,

$$e^{-t'/T_2} = \prod_i e^{-t_i/T_2}, \tag{7.22}$$

or in logarithmic form additively, making it applicable to cost vector analysis,

$$t' = \sum_i t_i, \tag{7.23}$$

providing a direct mechanism for calculating the effective dephasing rate across multiple subsequent sections in a qubit network route. The same can easily be seen to apply to depolarising, amplitude damping and loss channels.

Optical states

The notion of dephasing can be easily generalised to non-qubit states of light; i.e., with photon number $n > 1$. In general, dephasing has the property of mapping a superposition of basis states to a mixture of the same basis states, whilst preserving amplitudes. Thus, for perfect dephasing,

$$\mathcal{E}^{\text{dephasing}} \left(\sum_i \alpha_i |i\rangle \cdot \sum_j \alpha_j^* \langle j| \right) \rightarrow \sum_i |\alpha_i|^2 |i\rangle \langle i|, \qquad (7.24)$$

for some arbitrary basis enumerated by i and j. As an example, this process decoheres coherent states into thermal states. For partial dephasing, we can express the channel as creating a mixture over the input state with different phase rotations applied,

$$\mathcal{E}_\phi^{\text{dephasing}}(\hat{\rho}) = \int_0^{2\pi} \phi(\omega) \hat{\Phi}(\omega) \hat{\rho} \, \hat{\Phi}(\omega)^\dagger \, d\omega, \qquad (7.25)$$

where $\hat{\Phi}(\omega)$ is a phase-shift operator with phase ω, obeying $\hat{\Phi}(\omega)^\dagger = \hat{\Phi}(-\omega)$, and $\phi(\omega)$ is a normalised probability density function characterising the distribution of phase shifts. In the case of optical states, the phase-shift operators take the form

$$\hat{\Phi}(\omega) = e^{-i\omega\hat{n}} \qquad (7.26)$$

in the photon-number basis, where $\hat{n} = \hat{a}^\dagger \hat{a}$ is the photon-number operator, satisfying $\hat{n}|n\rangle = n|n\rangle$. With no dephasing, $\phi(\omega) = \delta(\omega)$ and $\mathcal{E}$ reduces to the identity channel. Otherwise, the off-diagonal (coherence) terms in the density operator begin to cancel out, leaving the diagonal (amplitude) terms unchanged. Thus, a perfect dephasing channel acting on a coherent state yields a thermal state of equal amplitude.

From this definition it can be seen that susceptibility to dephasing increases with photon number, because the number operator adds a multiplicative factor to the acquired phase shift,

$$\hat{\Phi}(\omega)|n\rangle = e^{-i\omega n}|n\rangle. \qquad (7.27)$$

For number states not in superposition, this corresponds to a simple unimportant global phase, because number states are phase-invariant. However, in superposition this adds relative phases, thereby destroying coherences upon applying the integral from Eq. (7.25).

7.3 Depolarisation

Depolarisation is a noise model more general than dephasing that probabilistically replaces a state with the completely mixed state (regardless of the input state). That is, with some probability we lose *all* quantum *and* classical information; i.e., both coherences and probability amplitudes. Note that the dephasing channel introduced above only destroys quantum coherence, whilst preserving amplitudes. Formally, the depolarising channel can be expressed as

$$\mathcal{E}_p^{\text{depolarising}}(\hat{\rho}) = p \cdot \hat{\rho} + (1 - p) \cdot \frac{\hat{I}}{\dim(\hat{\rho})}, \tag{7.28}$$

where $\hat{I}/\dim(\hat{\rho})$ is the completely mixed state in the d-dimensional Hilbert space.

When acting on qubits, the depolarising channel can equivalently be represented as the action of each of the four Pauli matrices with equal probability, because

$$\frac{\hat{I}}{2} = \frac{1}{4}(\hat{\rho} + \hat{X}\hat{\rho}\hat{X} + \hat{Y}\hat{\rho}\hat{Y} + \hat{Z}\hat{\rho}\hat{Z}). \tag{7.29}$$

Thus, both dephasing and depolarisation are examples of Pauli error models.

In the qubit basis (i.e., not including loss, for example), the Pauli matrices form a complete basis for quantum operations. Thus, the depolarising channel is the most general qubit error model, because it effectively applies all four Pauli error channels. For this reason, when evaluating fault-tolerance thresholds for QEC codes, thresholds are typically quoted in terms of the depolarising error rate.

Like the dephasing and loss channels, the error probability of multiple channels in series accumulates multiplicatively,

$$\mathcal{E}_{p_1}^{\text{depolarising}} \circ \mathcal{E}_{p_2}^{\text{depolarising}} = \mathcal{E}_{p_1 p_2}^{\text{depolarising}}. \tag{7.30}$$

7.4 Amplitude Damping

An error not so much relevant to optics but that arises very naturally in some other systems, such as atomic systems or quantum dots, is amplitude damping, also referred to as a *relaxation channel*. Here the process models the relaxation of a higher energy level, $|1\rangle$, to a lower energy one, $|0\rangle$. The $|0\rangle$ state is assumed to be the ground state and cannot relax any further, but the $|1\rangle$ state can spontaneously relax to the ground state. After complete amplitude damping, any input state will be left in the ground state $|0\rangle$. This model can be thought of as energy dissipating from the qubit system and being measured by the environment, leading to a type of decoherence whereby the input state is probabilistically replaced by the ground state.

The amplitude damping channel is easily represented in the quantum process formalism using two Kraus operators,

$$\hat{K}_1 = |0\rangle\langle 0| + \sqrt{\eta}|1\rangle\langle 1|,$$
$$\hat{K}_2 = \sqrt{1 - \eta}|0\rangle\langle 1|, \tag{7.31}$$

where $0 \leq \eta \leq 1$ quantifies the degree of damping ($\eta = 0$ represents complete damping, and $\eta = 1$ represents the identity channel).

The physical intuition is clear upon inspection of the structure of the projectors in the Kraus operators, with $\hat{K}_2$ representing relaxation from the excited state to the ground state, with probability $1 - \eta$.

In the specific context of optics, the loss channel is the equivalent of amplitude damping.

T_1-*Times*

The degree of amplitude damping is often quoted in terms of a channel's T_1-time, characterising the expected time for the excited state to undergo spontaneous emission and relax to the ground state. Using this parametrisation we can express the amplitude damping channel as

$$\mathcal{E}_t^{\text{relax}}(\hat{\rho}) = e^{-t/T_1}\hat{\rho} + (1 - e^{-t/T_1})|0\rangle\langle 0|, \tag{7.32}$$

for which the output state is of the form

$$\hat{\rho}_t = \begin{pmatrix} 1 - (1 - \alpha)e^{-t/T_1} & \gamma e^{-t/T_1} \\ \gamma^* e^{-t/T_1} & \beta e^{-t/T_1} \end{pmatrix}. \tag{7.33}$$

7.5 Mode-Mismatch

Mode-mismatch is an error model unique to optical implementations. For perfect interference to take place between two optical modes, which is necessary to entangle them or perform ideal 'which-path erasure',[3] the photons in those modes must be perfectly indistinguishable; i.e., they must exhibit identical spatiotemporal structure [148] and must be pure states.

---

[3] Which-path erasure is the phenomenon whereby a beamsplitter interaction between two modes makes processes associated with those two modes indistinguishable, thereby projecting them into a superposition state of both possibilities. This is most commonly used to entangle distinct photon-emitting systems.

This phenomenon arises very naturally whenever optical path lengths are not perfectly aligned or there is imperfect spatial mode overlap between optical modes interfering at beamsplitters. Furthermore, even if optical networks are perfect, photon distinguishability may arise during state preparation, because no two photon sources are absolutely identical – engineering photon sources is a precise business and no two are ever exactly alike.

In real-world experiments, the most common form of mode mismatch is temporal mode mismatch, whereby the timing of different photons is not perfectly synchronised, yielding temporal distinguishability, thereby reducing quantum interference. This type of error is easily introduced via mismatched path lengths in an experiment or incorrectly accounted for changes in refractive index. This is easily represented mathematically via translations in the temporal distribution functions of photons,

$$\psi(t) \rightarrow \psi(t - \Delta_t), \tag{7.34}$$

for temporal mismatch Δ_t. Of course, this logically generalises to other degrees of freedom, such as spatial mode mismatch, in which case a translation of the following form would take place:

$$\psi(x, y) \rightarrow \psi(x - \Delta_x, y - \Delta_y), \tag{7.35}$$

where x and y are the two transverse spatial dimensions perpendicular to the direction of propagation.

The Hong-Ou-Mandel (HOM) [88] *visibility* is a direct measure of the indistinguishability of two photons based on their interference fringes. Specifically, interference fringes are reduced as the photons become more distinguishable. Once completely distinguishable, they obey classical statistics.

Let us consider this in detail. Consider the two-mode, two-photon state

$$|\psi_{\text{in}}\rangle = \hat{A}^\dagger_{\psi_1} \hat{B}^\dagger_{\psi_2} |0\rangle, \tag{7.36}$$

where $\hat{A}^\dagger$ and $\hat{B}^\dagger$ denote the mode operators for two spatial modes, with respective temporal distribution functions ψ_1 and ψ_2. Evolving this through a 50:50 (Hadamard) beamsplitter yields

$$|\psi_{\text{out}}\rangle = \hat{U}|\psi_{\text{in}}\rangle \tag{7.37}$$

$$= \frac{1}{2}\left[\hat{A}^\dagger_{\psi_1} + \hat{B}^\dagger_{\psi_1}\right]\left[\hat{A}^\dagger_{\psi_2} - \hat{B}^\dagger_{\psi_2}\right]|0\rangle$$

$$= \frac{1}{2}\left[\hat{A}^\dagger_{\psi_1}\hat{A}^\dagger_{\psi_2} - \hat{A}^\dagger_{\psi_1}\hat{B}^\dagger_{\psi_2} + \hat{A}^\dagger_{\psi_2}\hat{B}^\dagger_{\psi_1} - \hat{B}^\dagger_{\psi_1}\hat{B}^\dagger_{\psi_2}\right]|0\rangle.$$

Post-selecting upon detecting a coincidence event (i.e., one photon per mode), the conditional state is projected onto

$$|\psi_{\text{cond}}\rangle = \frac{1}{2} \left[\hat{A}^\dagger_{\psi_1} \hat{B}^\dagger_{\psi_2} - \hat{A}^\dagger_{\psi_2} \hat{B}^\dagger_{\psi_1} \right] |0\rangle. \tag{7.38}$$

The probability of this coincidence event occurring is then given by the normalisation of the residual state,

$$P_{\text{coincidence}} = |\langle \psi_{\text{cond}} | \psi_{\text{cond}} \rangle|^2$$

$$= \frac{1}{2} - \frac{1}{2} \left| \int_{-\infty}^{\infty} \psi_1(t) \psi_2^*(t) \, dt \right|^2. \tag{7.39}$$

Now if we let both input photons have identical temporal structure, ψ, but with a time delay τ between them, this reduces to

$$P_{\text{coincidence}} = \frac{1}{2} - \frac{1}{2} \left| \int_{-\infty}^{\infty} \psi(t) \psi^*(t - \tau) \, dt \right|^2. \tag{7.40}$$

It is clear upon inspection that when $\tau = 0$, the coincidence probability $P = 0$, and we observe perfect photon bunching at the output (quantum statistics). On the other hand, as $\tau \to \pm\infty$, the photons become completely distinguishable, and we reduce to classical statistics, whereby $P = 1/2$. In the intermediate regime, there will be a monotonic trade-off between distinguishability (determined by $|\tau|$) and the coincidence probability. As an example, if we let the temporal distribution function be a normal Gaussian distribution,

$$\psi(t) = \frac{1}{\sqrt[4]{2\pi}} e^{-\frac{t^2}{4}}, \tag{7.41}$$

then

$$P_{\text{coincidence}} = \frac{1}{2} - \frac{1}{2} e^{-\frac{\tau^2}{8}}, \tag{7.42}$$

which is shown in Figure 7.4. Thus, experimentally measuring $P_{\text{coincidence}}$ directly determines the degree of photon distinguishability.

In the above representation of mode mismatch as a temporal or spatial translation, the process is entirely coherent and could in principle be reversed if the translation were known (which might easily be established using tomographic characterisation techniques). Of course, such translations could occur incoherently also. In particular, 'time jitter' is where this process occurs incoherently and the photons are subject to probabilistic temporal displacements. In this instance, a pure single-photon state would evolve into a mixture of states subject to

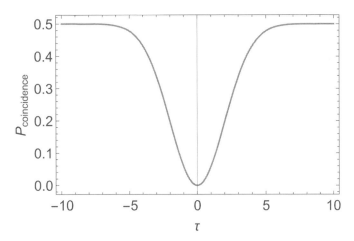

Figure 7.4 Hong-Ou-Mandel dip for two photons with normal Gaussian temporal distribution functions and temporal offset τ between them. τ effectively characterises the degree of photon distinguishability, where $\tau = 0$ represents complete indistinguishability (quantum statistics) and $\tau \to \pm\infty$ represents complete distinguishability (classical statistics). Thus, performing this experiment and measuring $P_{\text{coincidence}}$ can be used to characterise the degree of photon distinguishability.

different displacements. Because the mode mismatch is now probabilisitic, it is not reversible. The state of a single photon subject to time jitter would be of the form

$$\hat{\rho}_{\text{jitter}} = \int_{-\infty}^{\infty} p_{\text{jitter}}(\Delta_t) |\psi - \Delta_t\rangle\langle\psi - \Delta_t| d\Delta_t, \tag{7.43}$$

where $p_{\text{jitter}}(\Delta_t)$ characterises the classical probability distribution of the temporal displacement. Time jitter is particularly natural in heralded spontaneous parametric down-conversion (SPDC) sources, where imprecision in the measurement time of the heralding mode projects that temporal uncertainty onto the heralded state. For this reason, much time is being invested into engineering SPDC sources with separable output photons, such that pathological behaviour of the detection of the heralding photon does not project the heralded photon onto a mixed state. Time jitter is a major consideration in all present-day single-photon source technologies.

When considering mode mismatch, there are two general regimes for how it manifests itself in an optical system. The first is when the interference taking place is between distinct, independent photons; i.e., HOM interference (or its equivalent generalisations to higher-photon-number systems). The second is when multiple paths followed by a given photon interfere it with itself; i.e., MZ interference. The former only requires mode-matching on the scale of the photons' wave packets, whereas the latter requires interferometric stability on the order of the photons' wavelength, a far more demanding requirement.

Mode mismatch has been studied extensively in the context of linear optics quantum computing (LOQC). In particular, it was shown that in the cluster state formalism mode mismatch in a fusion gate is equivalent to a dephasing error model, where the dephasing rate is related to the degree of photon distinguishability (i.e., visibility) [151]. More generally, the operation of entangling gates [150, 149, 155, 152] and BOSONSAMPLING [143, 142] has been considered and explicit error models derived in the context of mode mismatch.

7.6 Dispersion

Dispersion is the phenomenon of frequency-dependent velocity of light in a given medium. These effects can be very diverse but can always be expressed in the mode-operator representation using an appropriate transformation in the temporal or spectral wave function,

$$f_{\text{disp}} : \tilde{\psi}(\omega) \rightarrow \tilde{\psi}(\omega)'. \tag{7.44}$$

7.7 Spectral Filtering

In Section 7.1 we discussed the loss channel, whereby with some fixed probability photons are lost to the environment. In reality, this process is often not uniform but frequency-dependent, resulting in spectral filtering effects. For example, optical fibres are typically designed to operate with a particular optical frequency in mind and will attenuate frequencies outside a given range, implementing, for example, low-pass, high-pass or bandpass spectral filtering.

Because spectral filtering can be regarded as frequency-dependent loss, it can be modelled in the same way as per the loss channel but using a frequency-dependent beamsplitter with transmissivity $\eta_f(\omega)$, which models the frequency response of the channel. The model is shown in Figure 7.5.

This channel has the effect of modulating the spectral distribution function of a photonic mode operator $\hat{A}_\psi^\dagger$ to $\hat{A}_{\psi'}^\dagger$, where

$$\psi'(\omega) = \sqrt{\eta_f(\omega)}\psi(\omega). \tag{7.45}$$

Note that unless $\eta_f(\omega) = 1 \ \forall \ \psi(\omega) \neq 0$, the new distribution function $\psi'(\omega)$ will not be normalised, where the normalisation reflects the loss probability

$$p_{\text{loss}} = 1 - \int_{-\infty}^{\infty} \eta_f(\omega)|\psi(\omega)|^2 \, d\omega. \tag{7.46}$$

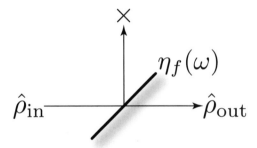

Figure 7.5 Model for the spectral filtering channel. The input state, $\hat{\rho}_{\text{in}}$, passes through a beamsplitter of frequency-dependent transmissivity $\eta_f(\omega)$ and the reflected mode is discarded, yielding the lossy output state $\hat{\rho}_{\text{out}} = \mathcal{E}_{\eta_f}^{\text{filter}}(\hat{\rho}_{\text{in}})$.

8

Quantum Cost Vector Analysis

As with the classical case in Section 2.2, there will be costs associated with the links and nodes in a network – nothing is free! In the quantum case, all of the usual classical costs are valid, but there are some very important additions of far greater relevance to most quantum applications. Classical digital data are discretised, resulting in data transmission highly robust against noise. In a quantum setting this is not necessarily the case, because the coefficients in quantum superpositions are continuous, meaning that errors accumulate during transmission and states will inevitably deteriorate, unlike digital states. This requires a rethinking of appropriate cost metrics.

8.1 Costs

We now briefly introduce some of the key measures for quantifying the quality of quantum communications links and how they may be expressed as metrics with meaningful operational interpretations. Many of the measures typically employed for characterising quantum systems are not true metrics (i.e., costs) but in many cases can be converted to metrics.

Efficiency

The efficiency measure introduced previously is multiplicative, so for consecutive lossy channels the net efficiency is

$$\eta_{\text{net}} = \prod_i \eta_i, \tag{8.1}$$

where η_i is the efficiency of the ith channel. Intuitively, this is simply telling us that if a photon passes through a channel with success probability η_1, followed by another with η_2, the total success probability is $\eta_1 \eta_2$.

When employing single-photon encoding of qubits (e.g., using the polarisation degree of freedom), there are three basis states of interest: a single photon horizontally polarised ($|H\rangle$), a single photon vertically polarised ($|V\rangle$) and the vacuum state ($|0\rangle$). The effect of the loss channel on this type of state is to map $|H\rangle$ and $|V\rangle$ to $|0\rangle$ with probability $1 - \eta$, while doing nothing to $|0\rangle$. Note that because the loss process affects both logical basis states ($|H\rangle$ and $|V\rangle$) identically, its action is invariant under unitary operations in the logical (i.e., polarisation) basis space.

Spectral Filtering

Because spectral filtering can be regarded as a frequency-dependent loss channel, its associated cost can be treated in the same manner, except that rather than keeping track of a single efficiency η, we track a frequency response function $\eta_f(\omega)$, with the same multiplicative property (on a per frequency basis),

$$\eta_f^{(\text{net})}(\omega) = \prod_i \eta_f^{(i)}(\omega). \tag{8.2}$$

If we are keeping track of the frequency response, the usual efficiency metric can be made redundant and absorbed into the frequency response function as a uniform response,

$$\eta_f(\omega) = \eta \; \forall \, \omega. \tag{8.3}$$

Decoherence

The dephasing and depolarising channels, given by Eqs. (7.13) and (7.28), also behave multiplicatively. If p_i is the probability that the state passing through the ith channel in series does not undergo the error process, then the probability of the state passing though the entire series without error is simply

$$p_{\text{net}} = \prod_i p_i, \tag{8.4}$$

exhibiting the same multiplicative behaviour as the loss channel. The same observation applies to any of the other Pauli error channels.

Mode Mismatch

In Section 7.5 we introduced a simple model for temporal mode mismatch as a displacement in the temporal wave function of photons propagating through a channel. Clearly, such a process is cumulative – a temporal displacement of Δ_1

followed by another of Δ_2 yields a net displacement of $\Delta_1 + \Delta_2$. Thus, for a chain of such channels we simply accumulate a net temporal displacement of

$$\Delta_{net} = \sum_i \Delta_i. \qquad (8.5)$$

For an incoherent mode-mismatching process, such as time jitter, an upper bound on the accumulated mismatch may be obtained by summing the maximum temporal displacements at each step.

Latency

Aside from the actual information content of a transmitted quantum state, the latency associated with its transmission is a key consideration in many time-critical applications.

By defining the latency of a link/node as the time between receipt of a quantum state and its retransmission, the total latency of a route is simply the sum of all of the individual node and link latencies across the route,

$$\mathcal{L}(R) = \sum_{i \in R} \mathcal{L}_i, \qquad (8.6)$$

where $\mathcal{L}_i$ is the latency associated with the ith link in route R.

Dollars

Not to be overlooked is the actual dollar cost of communicating information. It is unlikely that Alice and Bob outright own the entire infrastructure of particular routes. Rather, different links and nodes are likely to be owned by different operators (particularly in ad hoc networks), who are most likely going to charge users for bandwidth in their network (quantum networks will not be cheap). Clearly dollar costs are additive over the links and nodes within routes.

8.2 Costs as Distance Metrics

Definition 1 defines the properties of a cost metric in the classical context. We now wish to consider this in the quantum context, such that we are empowered to ask questions like 'What is the the total cost across a network route?' or 'Which route minimises a cost between two parties?', where *cost* now refers to some metric relevant to quantum state distribution, such as accumulated decoherence or loss.

If we consider a lossy photonic channel, for example, efficiencies (η) are multiplicative – for a route $v_1 \to v_2 \to v_3$, the net efficiency is given by the product of the individual efficiencies

$$\eta_{v_1 \to v_2 \to v_3} = \eta_{v_1 \to v_2} \eta_{v_2 \to v_3}. \tag{8.7}$$

This is multiplicative rather than additive, clearly not satisfying our definition for a cost metric. However, multiplicative metrics such as this can easily be made additive by shifting to a logarithmic scale, because

$$\log(\eta_{v_1 \to v_2 \to v_3}) = \log(\eta_{v_1 \to v_2}) + \log(\eta_{v_2 \to v_3}), \tag{8.8}$$

which now has a legitimate interpretation as a distance. The same applies to, for example, frequency response functions, which are equivalent to frequency-dependent loss.

In general, for a series of links $v_1 \to v_2 \to \cdots \to v_n$ characterised by multiplicative measure m, the equivalent cost metric is

$$c_{v_1 \to v_2 \to \cdots \to v_n} = -\sum_{i=1}^{n-1} \log(m_{v_i \to v_{i+1}}). \tag{8.9}$$

We have assumed that $0 \leq m \leq 1$, where $m = 0$ represents complete failure and $m = 1$ represents ideal operation.

With these properties, the costs in our graph have an elegant interpretation. In the case of perfect operation, $m = 1$, the cost is $c = 0$, creating an ideal direct link between neighbouring nodes at no cost. On the other hand, for complete failure, $m = 0$, the cost metric is $c = \infty$, effectively removing the link from the network and prohibiting path-finding algorithms from following that route altogether.

Such a logarithmic scale is particularly convenient when a cost metric over links accumulates on a per physical distance basis, in which case the cost metric is simply the physical length of the link multiplied by the metric per unit distance. For example, if a fibre channel implements loss at 3 dB/km, the loss over 10 km is 10×3 dB.

Note that lower bounds on fidelity, purity, efficiency and dephasing are all multiplicative on a scale of 0 to 1 and thus their logarithms may be regarded as cost metrics. Spatiotemporal mode mismatch, latency, dollar cost and displacements are clearly automatically metrics because they are additive.

A dephasing channel can be easily converted to a distance metric as follows. First we reparamterise the dephasing channel into

$$\mathcal{E}(\hat{\rho}) = p\hat{\rho} + (1 - p)\hat{Z}\hat{\rho}\hat{Z}$$
$$= (2p - 1)\hat{\rho} + (1 - p)(\hat{Z}\hat{\rho}\hat{Z} + \hat{\rho}). \tag{8.10}$$

Now $2p - 1$ is the probability that the state is not dephased and with the remainder replaced by the completely dephased state. Therefore, the probability of multiple applications of the channel $(\mathcal{E}_n \circ \cdots \circ \mathcal{E}_1)$ not dephasing the state scales multiplicatively as[1]

$$p_{\text{no error}} = \prod_i (2p_i - 1), \tag{8.11}$$

which is additive in a logarithmic scale as before,

$$\log(p_{\text{no error}}) = \sum_i \log(2p_i - 1), \tag{8.12}$$

which acts as a distance metric. This approach can similarly be applied to other Pauli channels.

In the case of mutual information and channel capacity, it makes most sense to consider the number of bits that are lost by a channel, rather than the number communicated, because then we have a measure with quasi-metric properties. Specifically, let the number of bits lost by a channel be the difference between the number of bits in the input state and the channel capacity,

$$B_{\text{lost}}(\mathcal{E}, \hat{\rho}) = S(\hat{\rho}) - C(\mathcal{E}). \tag{8.13}$$

Then there are two cases to consider – upper and lower bounds on accumulated lost bits.

The best-case scenario is that subsequent channels lose the same bits, giving us a lower bound on the number of lost bits as the maximum number of bits lost by the constituent channels,

$$B_{\text{lower}}(\mathcal{E}_2 \circ \mathcal{E}_1, \hat{\rho}) = \max[B_{\text{lost}}(\mathcal{E}_1, \hat{\rho}), B_{\text{lost}}(\mathcal{E}_2, \hat{\rho})]. \tag{8.14}$$

Alternately, each subsequent channel could lose a different set of bits, in which case the number of lost bits accumulates additively,

$$B_{\text{upper}}(\mathcal{E}_2 \circ \mathcal{E}_1, \hat{\rho}) = B_{\text{lost}}(\mathcal{E}_1, \hat{\rho}) + B_{\text{lost}}(\mathcal{E}_2, \hat{\rho}). \tag{8.15}$$

Then, the number of actual bits lost is bounded from above and below as

$$B_{\text{lower}} \le B_{\text{lost}} \le B_{\text{upper}}. \tag{8.16}$$

---

[1] With this parametrisation, in the limit of many applications of the dephasing channel, an input state asymptotes to the completely dephased state, $\lim_{n\to\infty} \mathcal{E}^n(\hat{\rho}) = \frac{1}{2}(\hat{Z}\hat{\rho}\hat{Z} + \hat{\rho})$. Thus, $2p - 1$ can be regarded as a discretised parametrisation of a system's T_2-time.

9

Routing Strategies

We introduced the notion of network costs in Section 2.2, strategies for allocating network resources in Subection 2.3 and a general formalism for optimising strategies minimise costs in Section 2.4. In this section we present some meaningful example strategies, illustrating the implementation of various aspects of strategies of practical real-world interest.

9.1 Single User

Let us begin our discussion of strategies by considering the simplest case of just a single user on the network. Consider the graph shown in Figure 2.2. This is the same example used earlier but now the edges have been weighted by some arbitrary cost metric. There are four routes from A to B. All have cost $c = 3$ except the route indicated by the red arrow, which has cost $c = 2$. Clearly the latter is optimal in terms of cost minimisation, and any shortest-path algorithm applied between A and B will accurately come to that conclusion. Thus, single-user networks are very trivial to optimise, and there is no distinction between LOCAL and GLOBAL strategies.

9.2 Multiple Users

Next consider the more complex network shown in Figure 9.1. We consider two sender/receiver pairs, $A_1 \rightarrow B_1$ and $A_2 \rightarrow B_2$. The available routes connecting both pairs overlap, creating competition for network resources.

Let us assume that there are just two properties of interest when deciding strategies – cost in dollars (which may differ for different links), augmented by their availability (i.e., whether a channel is currently available given the prospect of competition).

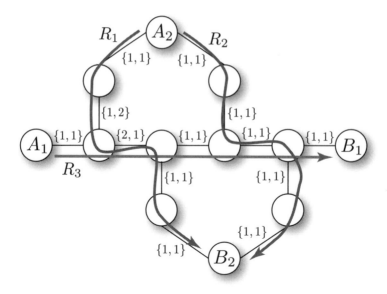

Figure 9.1 A simple network with two competing pairs of senders and receivers, $A_1 \to B_1$ and $A_2 \to B_2$. The optimal route for user B is dependent on competition from user A, illustrating the benefits of GLOBAL optimisation as opposed to individually based LOCAL optimisations.

Suppose the strategy optimises the $A_1 \to B_1$ route first, yielding R_3, before moving onto the $A_2 \to D_2$ route. In this case, the competing user implies the cheapest route R_2 is no longer available to be utilised simultaneously to R_3 and must therefore wait its turn on the following clock cycle. Alternatively, the strategy could employ R_1 for $A_2 \to B_2$, in which case their common link with capacity for two states would eliminate the competition between the two communications, allowing both to take place simultaneously. Thus, there is a trade-off: for $A_2 \to B_2$, we could achieve a net cost of $c(A_2 \to B_2) = 5$, requiring two clock cycles, or we could achieve simultaneous communication at the expense of increasing cost to $c(A_2 \to B_2) = 6$.

Suppose the $A_2 \to B_2$ route were optimised first. We would choose R_2. Then there would be a traffic jam on the $A_1 \to B_1$ route, and it would necessarily have to wait its turn. In a time-critical application, where waiting is intolerable, this effectively renders the network useless to the first sender/receiver pair.

If, however, the $A_1 \to B_1$ wa optimised first, choosing R_3, then R_2 would be prohibited under competition and the second best option, R_1, would be chosen. Now both communications could take place simultaneously. So we see that the outcomes of LOCAL optimisations need not always be consistent or unique. Rather, they can be highly dependent upon circumstantial issues, such as the arbitrary order in which routes are chosen for optimisation.

Generalising this to any number of users is a straightforward extension to the route optimisation problem, incurring a higher computational overhead due to the increased optimisation complexity.

In the upcoming sections we discuss multi-user strategies in more detail. None of these are true GLOBAL strategies but nonetheless address some of the problems facing LOCAL strategies mentioned above.

Truly GLOBAL strategies could employ, for example, the vehicle routing problem or vehicle rescheduling problem algorithms. However, both of these are **NP**-hard in general. Thus, the approximation heuristics to be discussed in the following sections are highly applicable.

10

Interconnecting and Interfacing Quantum Networks

Any global-scale network will inevitably comprise participants choosing to go about things their own way. The physical architecture and medium may vary from one subnetwork to the next, as may the network policies they adopt. The key then is to construct efficient *interconnects* between different levels of the network hierarchy, each of which may subscribe to their own local architecture-dependent network policies and cross between different physical mediums.

For example, the cost metrics employed at the intercontinental level would most certainly be very different to those in a small LAN. A small LAN might be running applications whereby they can easily reproduce packets and thereby tolerate packet loss. But for a warehouse-scale commercial quantum computing enterprise, responsible for performing one stage of a distributed quantum computation, the loss of a single packet could be extremely costly, requiring the entire computation to be performed completely from scratch due to no-cloning and no-measurement limitations, something that may not come cheaply.

Such interconnects will typically comprise a combination of the following:

- Packet switching: such that packets can be arbitrarily switched between the different levels of the network hierarchy.
- Physical interface: interconnect may be switching between different media. Such physical interfaces have costs associated with them. For example, coupling between free-space and fibre is typically very lossy.
- Quantum memory: such that data can be buffered while it awaits its turn at being switched between networks, because different networks may have different loads and operate at different clock rates.
- Packet format conversion: different levels of the network hierarchy may be employing entirely different cost functions, requiring packet headers to be reformatted upon switching between networks.

The packet switching and quantum memory are implemented as quantum processes at nodes, using the usual quantum process formalism. The physical interface between different mediums, if there is one, could be very diverse, encompassing many types of physical systems, but can always be characterised using the quantum process formalism. Packet headers, which contain all formatting, cost and routing information, are represented entirely classically and communicated entirely by the classical network. Thus, this operation also takes place at nodes but no quantum processes are taking place.

10.1 Optical Interfacing

Unless the entire pipeline of quantum operations through the course of a protocol is all-optical, there will be a need to exchange information between physical systems; for example, via light–matter interactions [47]. We will now discuss optical interfacing with some of the significant types of matter systems, such that their intercommunication can be optically mediated over the network.

Two-Level Systems

The archetypal interface is that between a photonic qubit in the $\{|0\rangle, |1\rangle\}$ photon-number basis and a two-level matter qubit in the $|g\rangle$ (ground) and $|e\rangle$ (excited) state basis. The logical qubit is defined as

$$|0\rangle_L \equiv |g\rangle,$$
$$|1\rangle_L \equiv |e\rangle. \tag{10.1}$$

Examples include atoms in cavities, NV centres and engineered quantum dots.

In the case of a photon interacting with a two-level matter qubit, the interface can be expressed via the Jaynes-Cummings interaction Hamiltonian of the form

$$\hat{H}_{\text{int}} = \hbar\chi(\hat{a}\,\hat{\sigma}^+ + \hat{a}^\dagger\hat{\sigma}^-), \tag{10.2}$$

where $\hat{a}$ ($\hat{a}^\dagger$) is the photonic annihilation (creation) operator, $\hat{\sigma}^\pm$ are the Pauli spin-flip operators and χ is the interaction strength. The interpretation of this Hamiltonian is very clear upon inspection – the annihilation (creation) of a photon is associated with the excitation (relaxation) of the two-level matter system, thereby directly coherently exchanging quantum information between the two systems, as shown in Figure 10.1.

λ-Configuration Systems

Alternatively, one can easily optically interface with a λ-configuration system, as shown in Figure 10.2. Here there are two degenerate ground states representing the

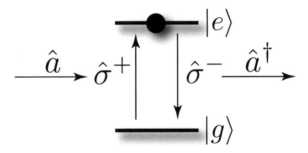

Figure 10.1 Light–matter interfacing between a single-photon state ($\hat{a}$, $\hat{a}^{\dagger}$) and a two-level matter qubit ($|g\rangle$, $|e\rangle$). The absorption (emission) of a photon is associated with the excitation (relaxation) of the matter qubit ($\hat{\sigma}^{\pm}$).

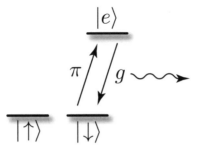

Figure 10.2 Light–matter interfacing between an optical mode and a λ-configuration system. The two degenerate ground states represent the logical qubit ($|0\rangle_L \equiv |\uparrow\rangle$, $|1\rangle_L \equiv |\downarrow\rangle$), only one of which may undergo transition to the excited state $|e\rangle$. Upon pumping the $|\downarrow\rangle \to |e\rangle$ transition with a π-pulse, a relaxation back to the ground state maps the logical qubit value to photon number.

logical qubit basis states ($|0\rangle_L \equiv |\uparrow\rangle$, $|1\rangle_L \equiv |\downarrow\rangle$), one of which may undergo a transition to an excited state, $|e\rangle$. By pumping the system to the excited state and waiting for a coherent relaxation, the emitted photon may be used to couple the qubit state of the λ-configuration to an optical mode, mapping the qubit value of the matter qubit to a photon-number representation.

Atomic Ensembles

In addition to single atoms with well-defined electronic structure, atomic ensembles [58, 44] can be used, whereby the absorption of a photon creates a *collective excitation* – a superposition of a single excitation across all atoms in the ensemble. Specifically, excitations are represented using collective excitation operators,

$$\hat{s}^{\dagger} = \frac{1}{\sqrt{N}} \sum_{i=1}^{N} \hat{s}_i^{\dagger}, \tag{10.3}$$

where

$$\hat{S}_i^\dagger = |e\rangle_i \langle g|_i \tag{10.4}$$

is the excitation operator for the ith atom in the ensemble, $|g\rangle_i$ and $|e\rangle_i$ are the ground and excited states for the ith particle and there are N atoms. The state of a single collective excitation is then given by

$$|\psi_{\text{collective}}\rangle = \hat{S}^\dagger |g\rangle^{\otimes N}. \tag{10.5}$$

Atomic ensembles are essentially well-engineered clouds of atomic gasses, trapped in a glass container, coupled to an optical mode. Atomic ensembles have been demonstrated with extremely long coherence lifetimes (T_2-times on the order of milliseconds), operating at room temperatures (a very attractive feature on its own). They exhibit *collective enhancement* in their coupling to the optical mode – the optical coupling strength is amplified by a factor quadratic in N compared to single-atom optical coupling, mitigating the need for a cavity.

The collective excitations exhibit the same general mathematical structure as single-atom excitations – the absorption (emission) of a single photon is associated with a single collective excitation (relaxation), albeit with the favourable collective enhancement in the coupling strength.

To couple an atomic ensemble (or other nonoptical system) with a polarisation-encoded photonic qubit, a PBS can be employed to spatially separate the horizontal and vertical modes, each of which couples to a separate atomic ensemble, which jointly represent the logical qubit, as shown in Figure 10.3.

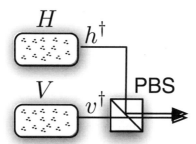

Figure 10.3 Coupling a polarisation-encoded photonic qubit to a pair of atomic ensembles, each of which corresponds to one of the qubit's logical basis states ($|0\rangle$ or $|1\rangle$). The horizontal and vertical components of the photonic qubit are spatially separated using a PBS and subsequently independently interface with distinct atomic ensembles via collective excitation.

Atomic ensembles have been proposed as quantum memories, given their long coherence lifetimes. Additionally, a protocol for universal cluster state quantum computation based on atomic ensemble qubits has been described [14].

Essentially, the long coherence lifetimes of collective excitations are due to the fact that the excitation is effectively encoded as a W-state, an equal superposition of a single excitation across many (N) atoms, of the form

$$
\begin{aligned}
|\psi_W^{(N)}\rangle = \frac{1}{\sqrt{N}}(&|e,g,g,\ldots\rangle \\
&+ |g,e,g,\ldots\rangle \\
&+ |g,g,e,\ldots\rangle \\
&+ \cdots \\
&+ |g,g,\ldots,e\rangle).
\end{aligned}
\tag{10.6}
$$

W-states are favourable from a decoherence perspective because tracing out a single particle has minimal impact on the coherence of the residual state, which preserves most entanglement, with this robustness growing with the number of particles. This is in stark contrast to Greenberger-Horne-Zeilinger (GHZ) states, which completely decohere under the loss of just a single particle.

Specifically, if $|\psi_W^{(N)}\rangle$ is the N-particle W-state (collective excitation), tracing out a single particle yields

$$
\begin{aligned}
\hat{\rho}_{tr} &= \mathrm{tr}_1(|\psi_W^{(N)}\rangle\langle\psi_W^{(N)}|) \\
&= \left(1 - \frac{1}{N}\right)|\psi_W^{(N-1)}\rangle\langle\psi_W^{(N-1)}| + \frac{1}{N}(|g\rangle\langle g|)^{\otimes(N-1)},
\end{aligned}
\tag{10.7}
$$

which for $N \gg 1$ approaches the pure state $|\psi_W^{(N-1)}\rangle$; i.e., a W-state with one fewer particles.

Superconducting Qubits

In the context of superconducting qubits, the energy difference between the energy levels being utilised to encode the qubit is extremely small. Therefore, photons coupled to these transitions sit in the microwave regime, whose wavelength lies in the range $\lambda \sim 100\,\mu\mathrm{m}$–$1\,\mathrm{m}$.

Information transfer between distinct superconducting qubits is achieved using a resonator, which acts as a quantum data bus. A simple resonator is an LC circuit, which can support only one frequency mode, but a waveguide resonator can support multiple modes. In general, the transmission line circuits used in nonlinear quantum electric circuits are in the form of coplanar waveguides. These waveguides are engineered to handle a particular set of frequencies and produce transmission lines with tunable frequency. Tunable resonators are very important in quantum optics and are useful in implementing controllable coupling between different quantum elements and also in shaping photon wave packets.

In cavity quantum electrodynamics (QED) the interaction of a natural atom with an optical photon in the visible wavelength regime is considered. Similarly, the interaction between quantum nonlinear electrical circuits and microwave photons is investigated in circuit QED. The coupling strength between a natural atom and visible light photon is fixed, where atoms couple weakly with photons [135]. Meanwhile, the coupling strength between a superconducting qubit and a microwave can be manipulated by engineering the parameters of the qubit and resonator, yielding strong and ultra-strong coupling between qubits and photons [186]. Furthermore, the coupling between an atom and a photon can be tuned dynamically during the course of an experiment. Several quantum optics components such as mirrors, beam splitters, circulators and switches can also be designed based on quantum electric circuits.

Due to scalability requirements, superconducting qubits are the most widely used qubit implementation used today. The energy-level spacing in superconducting qubits lie in the microwave regime. Hence, to control and transfer information from a superconducting qubit one might need to use microwave photons. In principle, we can use microwave photons to transfer information from one node to another, but such a transmission process is extremely lossy. Also, such processes have very demanding technical requirements, like the design of specialised Niobium waveguides, maintained at extremely low temperatures. Hence, it is not feasible to use microwave photons for the long-distance transfer of quantum information. It is well known that photons in the visible spectrum can be transmitted easily using optical fibres, with favourable efficiency.

To convert microwave photons to optical photons we can use a quantum transducer. A sketch of a typical design for a quantum transducer is shown in Figure 10.4 through a flowchart diagram. The quantum computer is made up of superconducting qubits, which feed quantum information to microwave qubits. The microwave qubits are then interfaced to optical qubits in the visible spectrum through a three-level quantum system, which can couple at both microwave and optical frequencies. These steps should be reversible. Hence, it should be possible to convert the optical qubits back into microwave qubits at the receiving end.

There are several proposals for quantum transducers in existence, and they can be classified into two major classes:

- Opto-mechanical [134, 16, 28, 54, 163, 167, 168].
- Spin ensembles [90, 27].

Figure 10.4 Block diagram for the quantum transducer. The quantum computer comprising superconducting qubits couples with the microwave qubit. The microwave qubit and the photonic qubit are coupled by a three-level system in which the energy difference between the levels corresponds to both microwave and visible optical frequencies. The thick arrows denote the forward process, required to convert the microwave to optical frequencies, and the dashed arrows represent the reverse process, required at the receiving end to convert the optical frequency to a microwave frequency.

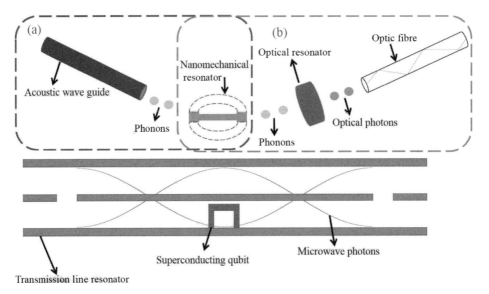

Figure 10.5 Scheme for an opto-mechanics-based quantum transducer. There are two possible ways of converting microwaves. The final steps for the process in which acoustic modes are transported using waveguides are enclosed by dashed box (a). Similarly, box (b) represents the scheme where phonons are converted to optical photons, which are then transmitted via optical fibres. The initial few steps consisting of the superconducting qubit, transmission line resonator and microwaves are common to both processes.

The opto-mechanical quantum transducer (see Figure 10.5), as the name suggests, combines optical components with a nano-mechanical resonator and converts the microwave photon into a phonon (acoustic) mode. The acoustic mode is then transmitted via waveguides. Because we need to fabricate waveguides with very high precision to transmit phonon modes, this scheme is not suitable for communicating between two distant quantum computers. A coupling of the phonon mode to the optical photon mode in the visible region was suggested to enable long-distance transfer of quantum information.

The Hamiltonian of an opto-mechanical quantum transducer, which converts a microwave photon to optical photon using an intermediate nano-mechanical resonator, reads

$$\hat{H} = \hbar\omega_1\,\hat{a}_1^\dagger\hat{a}_1 + \hbar\omega_2\,\hat{a}_2^\dagger\hat{a}_2 + \hbar\Omega\,\hat{b}^\dagger\hat{b} + \hbar\,g\,(\hat{b} + \hat{b}^\dagger)(\hat{a}_2^\dagger\hat{a}_1 + \hat{a}_1^\dagger\hat{a}_2), \qquad (10.8)$$

where ω_1 (ω_2) is the frequency of the microwave (optical) photon and Ω is phonon frequency. The operators $\hat{a}_1$ ($\hat{a}_1^\dagger$) and $\hat{a}_2$ ($\hat{a}_2^\dagger$) denote the annihilation (creation) operators corresponding to the microwave and optical photons, respectively. Meanwhile, $\hat{b}^\dagger$ ($\hat{b}$) denotes the phonon creation (annihilation) operator corresponding to the phonons. The factor g is the coupling strength between the microwave, phonon and photon modes. This design for a quantum transducer is widely preferred, because the optical photons in the visible spectrum can be transmitted over long distances using fibre optics. But the scheme requires two intermediate conversions, each of which reduces overall efficiency.

The spin ensemble–based quantum transducer (see Figure 10.6) is an alternative to the opto-mechanical quantum transducer. In this scheme an ensemble of spins interacts with microwave qubits via magnetic dipole coupling, and the superconducting qubits interact via electric dipole coupling with the microwave coupling. The Hamiltonian for such a system is

$$\hat{H} = \hat{H}_{\text{mw}} + \hat{H}_{\text{spin}} + \hat{H}_{\text{opt}}, \qquad (10.9)$$

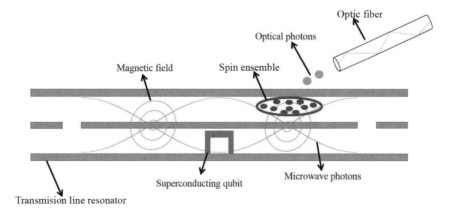

Figure 10.6 The spin ensemble–based quantum transducer. Microwave photons are converted to optical photons via a spin ensemble.

where

$$\hat{H}_{mw} = \hbar\omega_{sq}\hat{\sigma}_{sq}^{\dagger}\hat{\sigma}_{sq} + \hbar\omega_{\mu}\hat{a}^{\dagger}\hat{a} + \hbar g_{\mu}(\hat{a}^{\dagger}\hat{\sigma}_{sq} + \hat{a}\hat{\sigma}_{sq}^{\dagger}),$$
$$\hat{H}_{spin} = \hbar g_{s}(\hat{\sigma}_{ba}^{\dagger}\hat{\sigma}_{ba} + \hat{\sigma}_{bs}^{\dagger}\hat{\sigma}_{bs}),$$
$$\hat{H}_{opt} = \hbar g_{ab}(\hat{\sigma}_{ba}^{\dagger}\hat{c} + \hat{c}^{\dagger}\hat{\sigma}_{ba}). \tag{10.10}$$

The factor ω_{sq} is the frequency of the superconducting qubit, and $\hat{\sigma}_{sq}^{\dagger}$ and $\hat{\sigma}_{sq}$ are the raising and lowering operators corresponding to the superconducting qubits. The frequency of the microwaves is given by ω_{μ}, and the $\hat{a}^{\dagger}$ and $\hat{a}$ are the creation and annihilation operators for the microwave photons. The factor g_{s} is the coupling strength between the various levels of the spin, and $\hat{\sigma}_{ba}$ and $\hat{\sigma}_{bs}$ are the spin operators corresponding to the transition between the level a, b and s. Finally, the coupling strength of the spin interaction with the photon is denoted by g_{ab}, and $\hat{c}^{\dagger}$ ($\hat{c}$) is the creation (annihilation) operator corresponding to the photon. Again this is a two-step process, which is in addition beset with the problem of inhomogenous line broadening. The design of an experimental, high-fidelity quantum transducer is still an open and ongoing challenge in the field of quantum technology.

Part III

Protocols for the Quantum Internet

There are countless applications for the long-distance communication and processing of quantum data. We will outline some of the most notable examples. Broadly, we will begin with discussion of *low-level protocols* that form the primitives upon which other protocols are built. We will then progressively move towards *high-level protocols*, culminating with full *cloud quantum computing*.

Much of the recent experimental progress in quantum technology has been in the area of low-level protocols, although demonstrations of higher-level protocols are rapidly accelerating.

We keep in mind that although throughout this presentation we have been very quantum computing-centric, quantum computing is not the *only* quantum resource worth communicating. In the same way that *digital assets* encompass a broad range of digital systems and information, any aspect of a quantum system – from a state, to an operation, storage, measurement, or anything else – could be treated as a *quantum asset*, which, for generality, we would like our quantum networks to be able to handle.

At the lowest physical level, quantum protocols have in common that they involve state preparation, evolution and measurement as the fundamental primitives upon which more complex protocols are constructed. We consider these primitive resources in detail, before building upon them to consider some of the major elementary quantum protocols that implement tasks of practical interest. We treat full quantum computation separately in Chapter 27, because this is such an involved topic in its own right.

We will employ circuit model diagrams when describing some protocols. The unfamiliar reader may refer to Section 27.1 for a very brief introduction to quantum circuits.

Throughout this section the material will be optics-heavy and not include discussion of some purely nonoptical architectures, based on the reasonable assumption that networked quantum protocols will be optically mediated.

11

Optical Routers

Perhaps the most fundamental building block in any network is routers, devices that switch data packets between multiple inputs and outputs to relay them to a destination. Indeed, in many real-world networks, many nodes will purely implement routing and nothing more elaborate such as computations or other end-user protocols, to be discussed in Part III.

We now discuss the implementation of optical routers, beginning with the simplest two-port switch, upon which we build to construct more general and powerful routers.

There are many parameters of interest characterising the operation of optical routes. We will introduce the terminology convention:

- *Ports*: number of input and output optical modes in a device.
- *Channels*: number of simultaneous communications streams running in parallel through the device.
- *Optical depth*: number of primitive optical elements/devices an optical path traverses through the course of its trajectory from input to output.
- *Directionality*: whether information is transferred in one (unidirectional) or two (bidirectional) directions.
- *Switching time*: time for the switch to be reconfigured from one state to another.
- *Delay time*: time taken by a signal to reach the output line of the switch from input.
- *Throughput*: maximum data rate that can flow through the switch.
- *Switching energy*: energy input required for activating and deactivating the switch.
- *Power dissipation*: power dissipated during the process of switching.
- *Insertion loss*: loss in signal power when the switch is connected.
- *Crosstalk*: coupling to other optical modes.

Table 11.1 *Summary of different primitives for constructing optical routers. n, N_{bs}, N_{ps} and N_s are the number of input/output ports, beam splitters, phase-shifters and two-port switches, respectively. d is the optical depth (in units of number of two-port switches). Because all of the multiport devices are constructed from two-port switches, in all cases $N_{bs} = 2N_s$ and $N_{ps} = N_s$.*

Device	Resource requirements	Optical depth
Two-channel two-port	$N_{bs} = 2$, $N_{ps} = 1$	$d = 1$
Linear n-port	$N_s = n - 1$	$1 \leq d \leq n - 1$
Pyramid n-port	$N_s = n - 1$	$d = \log_2 n$
Single-channel multiport (linear)	$N_s = 2n - 3$	$2 \leq d \leq 2n - 3$
Single-channel multiport (pyramid)	$N_s = 2n - 3$	$d = 2 \log_2 n - 1$
Multi-channel multi-port	$N_s = \left\lceil \frac{n^2}{2} \right\rceil - n + 1$	$\left\lceil \frac{n}{2} \right\rceil \leq d \leq n - 1$
Crossbar	$N_s = n^2$	$1 \leq d \leq 2n - 1$

A summary of the routing devices we consider and their associated resource requirements is provided in Table 11.1.

Of course, real-world routers will not only switch optical paths but also implement some (probably undesired) quantum processes across those paths, such as a loss channel or temporal mode mismatch. Thus, proper analysis of optical router performance in quantum networks requires treating them as legitimate nodes in the network graph.

11.1 Mechanical Switches

Most obviously, optical switching could be performed mechanically, by physically displacing fibre endpoints, directing them towards different routes.[1] Such switches have found use in other areas but are not particularly appropriate for quantum information processing applications, because they are extremely slow compared to electro- or acousto-optic technologies. Certainly, mechanical switching would not be applicable to optical fast feedforward, such as that required by optical quantum computing, on the order of nanoseconds.

A second disadvantage of mechanical switches is that the introduction of moving parts into quantum optics protocols makes optical stabilisation extremely challenging. The mechanical control required to preserve wavelength-level coherence, for example, is effectively ruled out by moving mechanical parts.

---

[1] Remember, the telephone network used to be mechanically routed by human switchboard operators, manually routing point-to-point connections!

11.2 Interferometric Switches

Interferometric routers are based on the principle that the evolution implemented by interferometers is in general highly dependent on the phase relationships within them. This reduces the seemingly uphill task of high-speed, dynamic switching between modes to the problem of implementing dynamically controllable phases. Fortunately, there are a number of techniques for implementing such phase switching. We will discuss these phase modulation techniques before moving on to combining them into more complex routing systems.

A phase modulator is a classically controlled device that lets us tune the local phase accumulated by an optical path, ideally over the full range of $\{0, 2\pi\}$. These may be implemented in several ways.

Electro-optic Modulators

Electro-optic modulators (EOMs) are based on anisotropic materials, in which the refractive index changes according to an applied electric field. There are two primary variations on this:

- Pockels effect: a linear electro-optic effect, where the refractive index change is proportional to the applied electric field.
- Kerr effect: a quadratic electro-optic effect, where the refractive index change is proportional to the square of the applied electric field.

These changes in refractive index are typically small, such that the effects are significant over propagation distances larger than the light's wavelength. For example, in a material where the refractive index increases by 10^{-4}, an optical wave propagating a distance of 10^{-4} wavelengths will acquire a phase shift of 2π.

The refractive index of an electro-optic medium is a function $n(E)$ of the applied electric field E. This function varies only slightly with E, such that using a Taylor series expansion about $E = 0$ we obtain

$$n(E) = n + a_1 E + \frac{1}{2} a_2 E^2 + \cdots . \tag{11.1}$$

In a Pockels medium this relation becomes (after approximating and simplifying)

$$n(E) = n - \frac{1}{2} \chi n^3 E, \tag{11.2}$$

where χ is called the Pockels coefficient or linear electro-optic coefficient. Typical values of χ lie in the range $10^{-12} - 10^{-10} \mathrm{mV}^{-1}$. The most common crystals used as the medium for Pockels cells are $NH_4H_2PO_4$ (ADP), KH_2PO_4 (KDP), $LiNbO_3$, $LiTaO_3$ and CdTe.

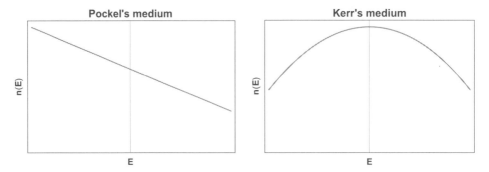

Figure 11.1 Dependence of refractive index on electric field in Pockels medium, exhibiting linear electric field dependence; Kerr medium, exhibiting quadratic electric field dependence. Graphs express qualitative behaviour only; hence, no numbers are provided.

In a centrosymmetric material or Kerr medium, this relation becomes (again after approximating and simplifying)

$$n(E) = n - \frac{1}{2}\xi n^3 E^2,$$ (11.3)

where ξ is the Kerr coefficient or the quadratic electro-optic coefficient. Typical values of ξ lie in the range $10^{-18} - 10^{-14} \mathrm{m^2 V^{-2}}$.

The refractive index profiles as a function of applied electric field strength for Kerr and Pockels mediums are shown in Figure 11.1.

Light transmitted through a transparent plate with controllable refractive index undergoes a controllable phase shift. This plate can be used as an optical phase modulator.

Consider light traversing a Pockels cell of length L to which an electric field E is applied. The phase shift undergone is given by

$$\phi \approx \phi_0 - \pi \frac{\chi n^3 E L}{\lambda_0},$$ (11.4)

where

$$\phi_0 = \frac{2\pi n L}{\lambda_0}.$$ (11.5)

If the electric field generated by applying a voltage V across the faces of the cell of dimension d is

$$E = \frac{V}{d},$$ (11.6)

then

$$\phi = \phi_0 - \pi \frac{V}{V_\pi}, \tag{11.7}$$

where

$$V_\pi = \frac{d\lambda_0}{L\chi n^3} \tag{11.8}$$

is the half-wave voltage, the voltage at which the phase shift changes by π.

The electric field is applied either perpendicular (transverse modulators) or parallel (longitudinal modulators) to the direction of the propagation of light. The value of the electro-optic coefficient χ depends on the directions of propagation and the applied field. The speed of operation is limited by the capacitive effects and the transit time of the signal through the material.

State-of-the-art electro-optic modulators are integrated optic devices based on LiNbO$_3$, in which materials like titanium are used to increase the refractive index. The typical operation speed is above 100 GHz. Light signals can be coupled in and out using optical fibres.

Acousto-optic Modulators

Sound, or acoustic, waves are vibrations that travel through a medium with a velocity characteristic of the medium. This can create perturbations in the refractive index of the optical medium, thus modifying the velocity of light passing through the medium. Thus, sound can be used to modify the effect of the medium on light. That is, sound can control the direction of propagation of light. This acousto-optic effect is used to make a variety of devices like optical modulators, switches, deflectors, filters, isolators, frequency shifters and spectrum analysers. This is shown in Figure 11.2.

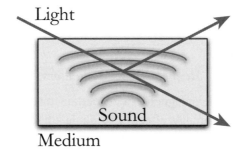

Figure 11.2 Acousto-optic modulators as a classically controlled optical switch. The light signal is refracted depending on the applied sound wave.

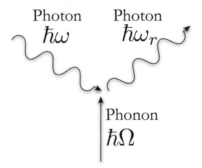

Figure 11.3 Energy diagram for an acousto-optic modulator, based on Eq. (11.9). Energy and momentum must be conserved from the incident photon of energy $\hbar\omega$ and phonon of energy $\hbar\Omega$, yielding a scattered photon of energy $\hbar\omega_r$.

According to quantum theory, a light wave of angular frequency ω and wave vector k is a stream of photons each with energy $\hbar\omega$ and angular momentum $\hbar k$. Additionally, acoustic waves with frequency Ω and wave vector q are a stream of phonons each with energy $\hbar\Omega$ and momentum $\hbar q$. When light and sound interact, a photon combines with a phonon to generate a new photon with energy and wave vector subject to energy and momentum conservation laws,

$$\hbar\omega_r = \hbar\omega + \hbar\Omega,$$
$$\hbar k_r = \hbar k + \hbar q. \tag{11.9}$$

The associated energy conservation diagram is shown in Figure 11.3.

Because the intensity of the reflected light is proportional to the intensity of the sound (provided the intensity of sound is low), the intensity of reflected light can be varied proportionally by using an electrically controlled acoustic transducer. This device can be used as a linear modulator of light.

When the acoustic power increases beyond a certain threshold level, total reflection of light occurs whereby the modulator behaves as an optical switch. By switching the sound on and off, the reflected light can be turned on and off, yielding an acoustically controlled switch.

Magneto-Optic Modulators

In the presence of a static magnetic field, certain materials act as polarisation rotators, known as the Faraday effect. The angle of rotation is proportional to distance and the rotary power ρ (angle per unit length), which is proportional to the component B of the magnetic flux density in the direction of wave propagation,

$$\rho = VB, \tag{11.10}$$

where V is known as the Verdet constant, which is a function of wavelength λ_0.

Examples of materials that exhibit the Faraday effect include glass, yttrium-iron-garnet (YIG), terbium-gallium-garnet (TGG) and terbium-aluminium-garnet (TbAlG).

A simple form of magneto-optic modulator comprises a parallel-sided disk of material placed in a small coil. An alternating current in the coil provides a magnetic field normal to the plane of the disk. The material becomes magnetised in this direction and light propagating through the disk undergoes a polarisation rotation about its plane of polarisation. The modulation of the angle of the plane of polarisation induced by the alternating current may be converted to amplitude modulation by subsequently passing the beam through a polariser.

11.3 Two-Channel Two-Port Switches

The elementary primitive switch from which more complicated routers may be constructed is the two-channel two-port switch. This switch may be constructed from a Mach-Zehnder (MZ) interferometer with a classically controlled phase shifter in one arm. By switching the phase to either $\phi = 0$ or $\phi = \pi$, the MZ may be tuned to implement either an identity or swap operation, respectively. This is shown in Figure 11.4.

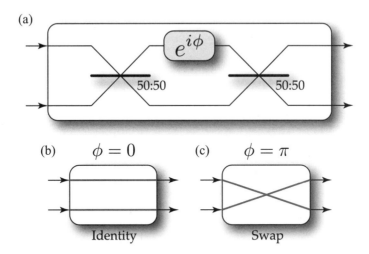

Figure 11.4 (a) A two-channel two-port switch has two inputs and two outputs, implementing either an identity or swap operation between them. This may be constructed using a Mach-Zehnder interferometer with a variable, classically controlled phase shift, $e^{i\phi}$, in one of the arms, which may be implemented using an acousto-optic or electro-optic modulator (AOM or EOM). The phase shift is allowed to be either $\phi = 0$ for an identity channel (b) or $\phi = \pi$ for a swap operation (c). Because the switch is based on MZ interference, this technique only applies to optical states that undergo MZ interference. The total resource requirements are two 50:50 beam splitters and a single phase shifter.

In the upcoming diagrams we present, arrows are used to indicate the time ordering of the flow of data. However, it should be noted that an MZ interferometer is reversible and therefore bidirectional, and so are all of the more complex routers based upon them.

Because the two-port switch is based on MZ interference, it will only function for optical states subject to such MZ interference. Thus, single photons and coherent states are applicable, whereas thermal states, for example, are not.

11.4 Multiplexers and Demultiplexers

From the two-port switch, which implements a controlled permutation of two optical modes, we can construct multiport multiplexers and demultiplexers, which controllably route a single input port to one of n multiple output ports or vice versa.

There are two main architectures that may be employed for implementing such multiplexers/demultiplexers. The first is to use a linear cascade of two-port switches, shown in Figure 11.5. The second is to use a pyramid cascade, shown in Figure 11.6. Both layouts require

$$N_s = n - 1 \tag{11.11}$$

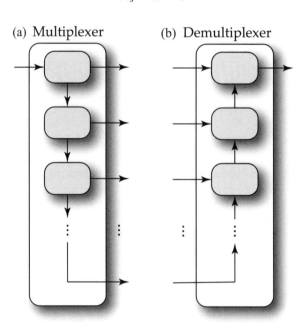

(a) Multiplexer   (b) Demultiplexer

Figure 11.5 Linear multiplexers (a) and demultiplexers (b) may be constructed from a linear chain of two-port switches (grey boxes), cascading into one another. These switch a single optical channel between n ports. The total resource requirements are $n - 1$ two-port switches. The optical depth ranges from 1 (for the first port) to $n - 1$ (for the final port).

(a)   Multiplexer

(b)   Demultiplexer

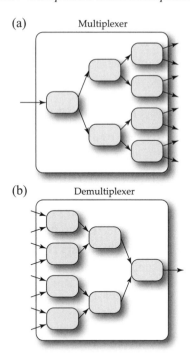

Figure 11.6 Pyramid multiplexers (a) and demultiplexers (b) decompose the multiplexing into a binary tree structure of two-port switches (grey boxes), shown here for the case of $n = 8$ ports. For n ports, all optical paths observe an optical depth of $d = \log_2(n)$ two-port switches, of which there are $n - 1$ in total.

two-port switches to implement. However, they differ in one important respect. In the linear multiplexer, different routes experience different optical depths, ranging from $d = 1$ (for the first port) to $d = n - 1$ (for the final port). This will lead to asymmetry in accumulated errors. In the pyramid multiplexer, on the other hand, all optical paths have the same optical depth, $d = \log_2 n$, yielding completely symmetric operation.

The differing optical depths of linear and pyramid multiplexers lend themselves naturally to different applications. Suppose that in a network a single input-to-output route through a multiplexer is used far more often than the others. In that case, utilising a linear multiplexer will minimise average optical depth because that route can be designated to the first output port, which has an optical depth of only $d = 1$. On the other hand, in a very balanced network, in which all optical routes are used roughly uniformly, the average case logarithmic optical depth of the pyramid multiplexer outperforms the average case linear optical depth of the linear multiplexer.

Note that the logarithmic optical depth of the pyramid configuration grows less quickly than the linear average optical depth of the linear configuration. Thus, on average, optical paths pass through fewer optical elements in the pyramid configuration, reducing average accumulated error rates when using noisy optical elements. This, in conjunction with the pyramid's perfect symmetry, makes the pyramid multiplexer configuration generally most favourable.

11.5 Single-Channel Multiport Switches

The multiplexers and demultiplexers route between one port and n ports. In the more general and useful case, we wish to route between n inputs and n outputs. If we only require one active channel at a given time, such a router may be trivially constructed from an n-port multiplexer connected to and n-port demultiplexer, as shown in Figure 11.7. Here, the demultiplexer chooses one of the input modes to route to its single output, which then feeds into the multiplexer to fan it out to the desired output. The multiplexers/demultiplexers could be implemented using either of the aforementioned layouts, yielding a total resource count of

$$N_{\mathrm{s}} = 2n - 3 \tag{11.12}$$

two-port switches.[2]

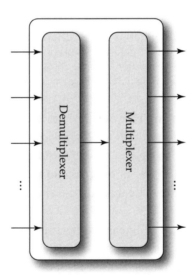

Figure 11.7 A single-channel multiport switch may be constructed by demultiplexing the n input ports to a single port and routing the desired input channel to that port before multiplexing it back out to the desired output port. This allows an arbitrary input to be routed to an arbitrary output but only one channel at a time. This requires $2n - 3$ two-port switches in total.

---

[2] Note that the multiplexer and demultiplexer each require $2(n - 1)$ two-port switches, but one of the central ones adjoining the multiplexer and demultiplexer is redundant and may be eliminated, reducing the number of two-port switches to $2n - 3$.

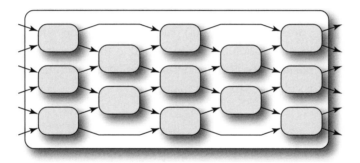

Figure 11.8 A completely general multichannel multiport switch may be constructed using a staggered grid of two-port switches (grey boxes), shown here for $n = 6$ ports. This allows the implementation of an arbitrary permutation between input and output ports, enabling all n channels to be simultaneously utilised and routed across distinct input-to-output routes. This requires $\left\lceil \frac{n^2}{2} \right\rceil - n + 1$ two-port switches in total. Optical depth is approximately equal across all input-to-output paths.

11.6 Multichannel Multiport Switches

The single-channel multiport switch enables switching between an arbitrary number of input/output ports but it can only route a single channel at a time. The most general scenario to consider is multichannel multiport switching, which implements an arbitrary permutation between n inputs and n outputs. That is, all n ports may be routing active channels, enabling simultaneous routing of multiple data flows.

Such a switch may be constructed from a staggered, rectangular lattice of two-port switches, as shown in Figure 11.8. It is easy to see upon inspection that optical paths exist between every input/output pair of ports. The total resource count for this device is

$$N_s = \left\lceil \frac{n^2}{2} \right\rceil - n + 1 \tag{11.13}$$

two-port switches.

The operation implemented by this device can therefore be expressed as

$$\begin{pmatrix} \hat{b}_1^\dagger \\ \hat{b}_2^\dagger \\ \vdots \\ \hat{b}_m^\dagger \end{pmatrix} = \hat{\sigma} \cdot \begin{pmatrix} \hat{a}_1^\dagger \\ \hat{a}_2^\dagger \\ \vdots \\ \hat{a}_m^\dagger \end{pmatrix}, \tag{11.14}$$

where $\hat{\sigma} \in S_m$ is an arbitrary element of the symmetric group (i.e., a permutation matrix) and $\hat{a}_i^\dagger$ ($\hat{b}_i^\dagger$) are the input (output) photonic creation operators.

Note that this decomposition is more favourable than the completely general Reck *et al.* decomposition because the circuit is balanced, with (almost!) identical optical depths across all input-to-output paths.

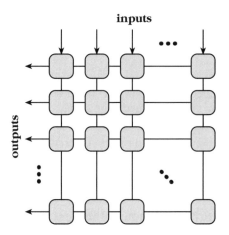

Figure 11.9 Crossbar architecture for multiport switching. Each box represents a 2×2 switch, of any physical implementation. The switching sequence of the constituent two-port switches is defined by a binary $n \times n$ permutation matrix, whose elements determine whether a given two-port switch flips modes or not.

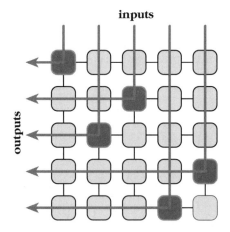

Figure 11.10 Example switching configuration for a 5×5 crossbar switch. The switch shading represents whether the respective 2×2 switch is set to flip (dark) the modes or not (light).

11.7 Crossbar Switches

A general multiport switching architecture, which gained popularity in the early days of channel-switched telecommunications networks, is the crossbar architecture, whereby n inputs are mapped to n outputs via a binary permutation matrix, which controls a lattice of 2×2 switches. The general layout of the architecture is shown in Figure 11.9, and an example of a routing sequence corresponding to a particular binary control matrix is shown in Figure 11.10.

Clearly the scheme requires n^2 two-port switches to implement arbitrary $n \times n$ mode permutations. The main disadvantage of this scheme is that in general different paths within a given permutation experience differing optical depths, ranging from 1 (best case) to $2n - 1$ (worst case).

12

Optical Stability in Quantum Networks

Given that communications links in quantum networks are expected to be optical, an issue of central importance is optical stability when signals from remote sources interfere or interact with local quantum states. For example, in an entanglement swapping protocol forming a part of a quantum repeater network, if the entangling operation between the remotely prepared qubits suffers errors, so, too, will the prepared distributed entangled state.

If we consider the simplest scenario of employing a polarising beam splitter to implement the entangling operation in the polarisation degree of freedom, photon distinguishability in the form of mode mismatch will undermine quantum interference, thereby reducing the entangling power of the gate. Similar observations apply to many other protocols involving entangling measurements or multiphoton interference more generally.

In present-day laboratories, mode mismatch and photon distinguishability can be controlled with exceptionally high fidelity. However, in the networking context this is likely to not be so easy, because perfectly aligning states emanating over long-distance communications channels, which we do not have exquisite control over in a well-controlled laboratory setting, is going to be a somewhat unpredictable and time-varying technological challenge.

Such processes are likely to arise in a multitude of ways, including, but certainly not limited to the following:

- Optical fibre: slight variations in temperatures induce refractive index changes, or changes in physical dimension, resulting in temporal displacements of optical wave packets.
- Satellite: precise knowledge of the distance to a rapidly moving target, at the scale of photon wave packets, is an extremely daunting prospect.
- Free-space (including via satellite): unpredictable temperature and pressure fluctuations in the atmosphere cause unpredictable variations in the speed of light.

For these inevitable reasons, it is important to understand the susceptibility of different network protocols to optical stability.

There are two dominant forms of photonic interference that must be considered, each with quite distinct behaviours under the influence of optical instability. These are the following:

- Hong-Ou-Mandel (HOM) interference: interference between two distinct photons at a beam splitter.
- Mach-Zehnder (MZ) interference: self-interference of a single-photon traversing multiple paths in superposition within an interferometer.

12.1 Photon Wave Packets

Before describing optical interference in detail, we must first formalise a definition for the optical wave packets we will be dealing with. We will assume wave packets with Gaussian temporal envelope of width σ (the coherence length), frequency shifted by some carrier frequency ω_0 (the wavelength).

The temporal distribution function is then

$$\psi(t) = \sqrt[4]{\frac{2}{\sigma\pi}}e^{-\frac{t^2}{\sigma}-i\omega_0 t}, \tag{12.1}$$

with associated mode operator $\hat{A}_\psi^\dagger$. This wave packet is normalised such that

$$|\langle 0|\hat{A}_\psi \hat{A}_\psi^\dagger|0\rangle|^2 = \int_{-\infty}^{\infty} |\psi(t)|^2 \, dt = 1. \tag{12.2}$$

Of course, the temporal envelope need not be Gaussian in general and could take any other form, subject to normalisation. In Figure 12.1 we illustrate the two main features of this representation: the temporal envelope and the underlying carrier frequency that it modulates.

In real-world scenarios we are likely to encounter carrier frequencies sufficiently large that oscillations at the carrier frequency level are far more rapid than that of the temporal envelope. For this simple reason, it is to be expected that interference dependent only on σ will be far more robust against temporal instability than interference dependent on ω_0.

12.2 Mach-Zehnder Interference

MZ interference is the interference of a photon or coherent state with itself in a two-mode interferometer constructed from two 50:50 beam splitters in series, as shown in Figure 12.2(a). This is MZ interference in its simplest form, which can,

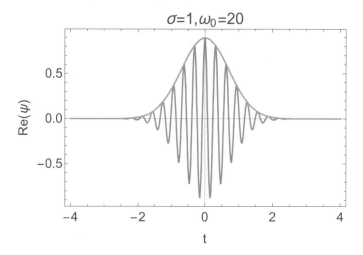

Figure 12.1 A photonic wave packet of the form of Eq. (12.1), with Gaussian temporal envelope of width σ, shifted by a much higher carrier frequency ω_0.

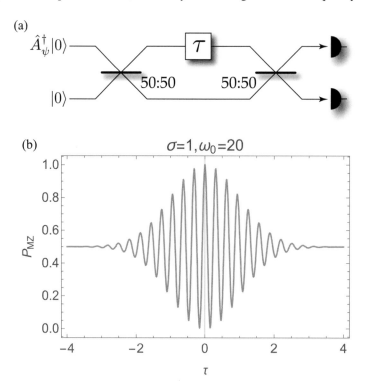

Figure 12.2 Mach-Zehnder self-interference of a single photon. (a) Layout of the interferometer. The photon is subject to a time delay τ within only the upper arm of a balanced interferometer, comprising two 50:50 beam splitters. (b) Interference fringe $P_{MZ}(\tau)$ as a function of the time delay. The interference is sensitive at the scale of the photon wavelength.

of course, be generalised to more complex networks involving self-interference across multiple optical paths.

Within the interferometer is a time delay, τ, which acts as a temporal mismatch between the two optical paths.

Let us calculate explicitly the evolution of a single photon through this device, beginning with a photon described by mode operator $\hat{A}_\psi^\dagger$, with the temporal distribution function from Eq. (12.1). We have

$$
\begin{aligned}
|\psi_{\text{in}}\rangle &= \hat{A}_\psi^\dagger |0,0\rangle \\
&\xrightarrow{\text{BS}} \frac{1}{\sqrt{2}}[\hat{A}_\psi^\dagger + \hat{B}_\psi^\dagger]|0,0\rangle \\
&\xrightarrow{\tau} \frac{1}{\sqrt{2}}[\hat{A}_{\psi-\tau}^\dagger + \hat{B}_\psi^\dagger]|0,0\rangle \\
&\xrightarrow{\text{BS}} \frac{1}{2}[\hat{A}_{\psi-\tau}^\dagger + \hat{B}_{\psi-\tau}^\dagger + \hat{A}_\psi^\dagger - \hat{B}_\psi^\dagger]|0,0\rangle \\
&\xrightarrow{\text{PS}} \frac{1}{2}[\hat{A}_{\psi-\tau}^\dagger + \hat{A}_\psi^\dagger]|0,0\rangle \\
&= \frac{1}{2}\int_{-\infty}^{\infty} [\psi(t) + \psi(t-\tau)]\hat{a}^\dagger(t)\,dt,
\end{aligned}
\tag{12.3}
$$

where BS denotes the evolution implemented by a 50:50 beam splitter, and PS denotes post-selecting upon detecting a single photon in the first output mode.

We now characterise the operation of the device in terms of the probability of detecting the photon in the first output mode,

$$
\begin{aligned}
P_{\text{MZ}}(\tau) &= \frac{1}{4}\int_{-\infty}^{\infty} |\psi(t) + \psi(t-\tau)|^2\,dt \\
&= \frac{1}{2}\left[1 + e^{-\frac{\tau^2}{2\sigma}}\cos(\omega_0\tau)\right].
\end{aligned}
\tag{12.4}
$$

These dynamics are shown in Figure 12.2(b). There are two key features in the behaviour of $P_{\text{MZ}}(\tau)$. First, there is a slowly varying Gaussian term. Second, the Gaussian term modulates a rapidly oscillating sinusoidial term associated with the carrier frequency. This implies that τ on the order of the photon's wavelength dominates the measurement dynamics, making it extremely sensitive to temporal instability.

12.3 Hong-Ou-Mandel Interference

In HOM interference, there is no self-interference as per MZ but rather interference between two independent but indistinguishable photons. The interference takes

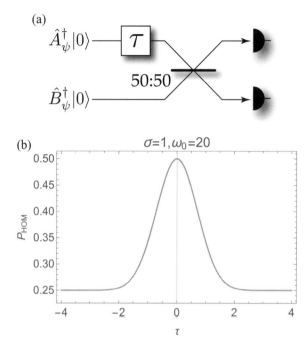

Figure 12.3 Hong-Ou-Mandel interference between two independent photons, A and B. (a) Layout of the interferometer. Two photons given by mode operators $\hat{A}^{\dagger}$ and $\hat{B}^{\dagger}$, both with temporal distribution function $\psi(t)$, interfere at a 50:50 beam splitter, where mode A is first subject to a time delay τ. (b) Interference fringe $P_{\text{HOM}}(\tau)$ as a function of the time delay. The fringe is only sensitive at the scale of the wave packet envelope (outer curve) in Figure 12.1.

place at a single 50:50 beam splitter, with a temporal delay in one input mode modelling temporal instability. The model is shown in Figure 12.3(a).

Performing the same evaluation of the evolution of the system as before, we obtain

$$
\begin{aligned}
|\psi_{\text{in}}\rangle &= \hat{A}_{\psi}^{\dagger}\hat{B}_{\psi}^{\dagger}|0,0\rangle \\
&\xrightarrow{\tau} \hat{A}_{\psi-\tau}^{\dagger}\hat{B}_{\psi}^{\dagger}|0,0\rangle \\
&\xrightarrow{\text{BS}} \frac{1}{2}[\hat{A}_{\psi-\tau}^{\dagger} + \hat{B}_{\psi-\tau}^{\dagger}][\hat{A}_{\psi}^{\dagger} - \hat{B}_{\psi}^{\dagger}]|0,0\rangle \\
&\xrightarrow{\text{PS}} \frac{1}{2}\hat{A}_{\psi}^{\dagger}\hat{A}_{\psi-\tau}^{\dagger}|0,0\rangle \\
&= \frac{1}{2}\int_{-\infty}^{\infty}\int_{-\infty}^{\infty}\psi(t)\psi(t'-\tau)\hat{a}^{\dagger}(t)\hat{a}^{\dagger}(t')\,dt\,dt'|0,0\rangle.
\end{aligned}
\tag{12.5}
$$

We then characterise the operation of the device in terms of the probability of detecting both photons in the first output mode (photon bunching),

$$P_{\text{HOM}}(\tau) = \frac{1}{4} \left[1 + \left| \int_{-\infty}^{\infty} \psi(t)\psi(t-\tau)^* \, dt \right|^2 \right]$$

$$= \frac{1}{4} \left[1 + e^{-\frac{\tau^2}{\sigma}} \right]. \tag{12.6}$$

These dynamics are shown in Figure 12.3(b). Now, unlike MZ interference, we observe no dependence on the carrier frequency and its associated rapidly oscillating terms. Rather, operation depends only on the temporal envelope, which exists over a far larger timescale.

Importantly, unlike MZ interference, HOM interference is not applicable to coherent states, which do not entangle or enter into superposition at beam splitters. The photon bunching effect is unique to single photons.

The intuition behind the HOM-dip phenomenon is as follows. We know that for identical, indistinguishable photons, an input photon pair evolves as

$$|1,1\rangle \underset{\text{BS}}{\rightarrow} \frac{1}{\sqrt{2}}(|2,0\rangle - |0,2\rangle), \tag{12.7}$$

yielding perfect photon bunching. This bunching effect arises from quantum mechanical interference between the photons. Next, imagine that the two photons arrived a long time apart from one another, so long that their wave packets do not overlap at all. In that instance, the photons do not 'see' one another and no quantum interference takes place. Instead, rather than a two-photon quantum interference experiment, we effectively have two independent instances of single-photon experiments, given by

$$|1,0\rangle \underset{\text{BS}}{\rightarrow} \frac{1}{\sqrt{2}}(|1,0\rangle + |0,1\rangle),$$

$$|0,1\rangle \underset{\text{BS}}{\rightarrow} \frac{1}{\sqrt{2}}(|1,0\rangle - |0,1\rangle). \tag{12.8}$$

Note that each of these independent instances obeys the classical statistics of a 50:50 distribution. Combining the two instances using classical probability theory, we now observe a 50% chance of measuring a coincidence, as opposed to the 0% chance for true HOM interference.

12.4 HOM vs MZ Interference

Let us now examine the implications of these different types of interference. The key observation was that MZ is far more sensitive to temporal mismatch than HOM, the former at the scale of the photons' wavelength and the latter at the scale of their temporal envelope, which is far larger.

This leads to the immediate conclusion that network protocols relying on HOM interference will be far more robust against temporal instability than those relying on MZ interference. Realistically, it is to be expected that the latter might be impossibly challenging in many contexts, because wavelength-scale stabilisation over long distances seems implausible.

13

State Preparation

The first step in any quantum protocol involves the preparation of some kind of quantum state. Some quantum states are easy and cheap to prepare. Others are complex and costly. Thus, the most fundamental quantum asset that a quantum network must handle is the preparation and communication of quantum states.

A state prepared by Bob and sent to Alice might be prepared in isolation, or it might be entangled with a much larger system held by Charlie that Alice does not have full access to. In that case, it would be impossible for Alice to prepare the state on her own, unless she were to first establish a relationship with Charlie. Alternatively, maybe Alice just is not very well resourced, and cannot do much on her own. The ability to let someone else prepare her desired quantum states for her would be highly appreciated.

Given the emphasis on quantum optics in quantum networking, it should be noted that optical quantum state engineering has broad applications but can be very challenging in general. Single-photon state engineering, for example, finds ubiquitous applications in quantum information processing protocols and has become commonplace. Most notable, linear optics quantum computing and some quantum metrology protocols rely on single-photon state preparation. 'Push-button' (i.e., on-demand) single-photon sources would be a prized asset to many undergraduate experimentalists, were they able to afford them. But with access to the quantum internet, they could purchase single photons from another better-resourced lab, with quality of service constraints guaranteed by the quantum network protocol.

Even the most basic primitive in quantum technologies – state preparation – already brings with it much to take into consideration when designing quantum networks, which future quantum network stacks (codename QTCP) should accommodate.

13.1 Coherent States

Coherent states, although not strictly *quantum* states, nonetheless find broad applications in quantum protocols; for example, as the pump for SPDC sources or as a phase reference for homodyne detection. Coherent states are rather trivial to prepare, because they are closely approximated by laser sources. Despite their triviality, high-quality lasers can nonetheless become very expensive, large and inaccessible to the not-so-well-resourced end-user. It is not uncommon for laser sources in contemporary labs to be valued in the $100Ks.

13.2 Single Photons

Single-photon sources [129] are of particular interest, as a foundational building block in many optical quantum information processing applications, such as linear optics quantum computing and quantum key distribution.

The most common approach to preparing single-photon states is via heralded SPDC [178, 179], whereby a coherent pump source is down-converted into two-mode photon pairs via a second-order nonlinear crystal with interaction Hamiltonian of the form

$$\hat{H}_{\text{SPDC}} = \xi (\hat{a}_p \hat{a}_s^\dagger \hat{a}_i^\dagger + \hat{a}_p^\dagger \hat{a}_s \hat{a}_i), \tag{13.1}$$

where ξ is the interaction strength and $\hat{a}_p$, $\hat{a}_s$ and $\hat{a}_i$ are the photonic annihilation operators for the pump (input) and *signal* and *idler* (output) modes, respectively. This has the clear intuitive interpretation as the coherent exchange of photon pairs in the output modes with photons in the coherent pump.

Specifically, a two-mode SPDC state takes the form

$$|\psi\rangle_{\text{SPDC}} = \sqrt{1 - \chi^2} \sum_{n=0}^{\infty} \chi^n |n\rangle_s |n\rangle_i, \tag{13.2}$$

where χ is the squeezing parameter, a function of the pump power and properties of the crystal. The layout is shown in Figure 13.1.

Applying the single-photon projector, $|1\rangle\langle 1|$, to the first mode yields the single-photon state in the other, up to normalisation, which reflects the inherent nondeterminism. The preparation success probability is derived from the amplitude of the $n = 1$ term as

$$P_{\text{prep}} = \chi^2 (1 - \chi^2), \tag{13.3}$$

assuming ideal photodetection. Thus, the perfect photon number correlation enables heralded preparation of states with exactly one photon in principle.

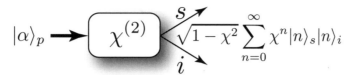

Figure 13.1 Layout of an SPDC single-photon source. A second-order nonlinear crystal is pumped with a coherent state (i.e., laser source), yielding a two-mode output state with perfect photon number correlation between the two modes. Then, post-selecting upon detecting a single photon in one mode in principle guarantees a single photon in the other.

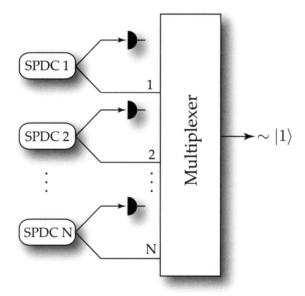

Figure 13.2 Quasi-deterministic single-photon state preparation using N-fold multiplexing of heralded SPDC sources (or any other nondeterministic but heralded source). All N SPDC sources are triggered simultaneously. The heralding detectors feedforward to the multiplexer, which routes a successfully heralded single-photon state (if there is one) to the output mode. With a sufficiently large bank of sources in parallel, the probability of successfully preparing a single-photon state approaches unity.

Transitioning from heralded state preparation to quasi-deterministic state preparation may then be achieved by operating a bank of such sources in parallel and multiplexing their outputs such that, when all sources are triggered simultaneously, if any one succeeds the respective single photon is routed to the desired output mode, as shown in Figure 13.2.

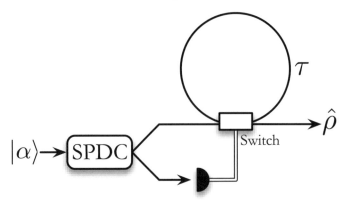

Figure 13.3 Multiplexed single-photon state preparation in the temporal domain. An SPDC source operating at high repetition rate, with time bin separation τ, enters a fibre loop with an in/out coupling switch classically controlled by the heralding outcomes. The fibre loop acts as a quantum memory, keeping the most recent successfully heralded time bin in memory until the procedure terminates. The output is a pulse train where the last time bin closely approximates a single photon.

The success probability of the multiplexed source exponentially asymptotes to unity as the number of in-parallel sources increases,

$$P_{success} = 1 - (1 - P_{prep})^N, \tag{13.4}$$

where there are N sources in parallel. This principle could also obviously be applied to any other type of nondeterministic but heralded source.

The multiplexing approach need not be restricted to the spatial domain but could also be equivalently implemented in the temporal domain [147].

Of course, operating a large bank of sources in parallel, along with the associated multiplexing, which requires nanosecond-scale fast feedforward, is experimentally costly (in physical size, complexity and dollars), making outsourcing of this technology potentially highly desirable.

This description is purely in the photon number basis. However, as discussed in Section 6.3, photons also have spatiotemporal characteristics. This strongly affects state preparation when using heralded SPDC, particularly state purity, and much effort has been invested into engineering the spectral structure of SPDC states to maximise purity and indistinguishability [5, 30]. Specifically, we wish to engineer the photon pairs to be spectrally separable, such that the heralded photon remains spectrally pure even if the heralding photon was measured with undesirable spectral characteristics (e.g., finite resolution).

SPDC is relatively cheap and widely used but nonetheless might be out of reach for many end-users, particularly when the previously discussed multiplexing techniques are employed to boost heralding efficiencies. It is quickly being superseded by superior technologies in cutting-edge labs, such as quantum dot sources, which have deterministic, push-button potential [159, 100]. Techniques based on cavity quantum electrodynamics (QED) [31] and molecular fluorescence [38] have also been demonstrated. However, such sources are very much in their developmental stages and relatively expensive.

Generally speaking, a push-button photon source could be constructed from any two-level system, comprising a ground state, $|g\rangle$, and an excited state, $|e\rangle$, with short lifetime, whereby relaxation via the $|e\rangle \rightarrow |g\rangle$ transition emits a photon. Then, pumping the system to excite it to the $|e\rangle$ state and waiting for spontaneous decay yields a single photon.

13.3 Cluster States

In addition to the simple single- or two-mode states discussed above, an entire universal quantum computation can be performed using the 'cluster state' measurement-based model for quantum computation (explained in detail in Section 27.2). Here state preparation can not only be outsourced but distributed, with different hosts preparing different geometric parts of the state, which are then 'stitched together'.

The beauty of this type of state is that there is a natural separation between state preparation and computation, with the preparation stage being far more technologically challenging than the computation stage. Thus, Alice might ask better-resourced Bob to prepare a cluster state and send it to her, at which point she implements the computation herself using only simple single-qubit measurement operations.

13.4 Greenberger-Horne-Zeilinger States

Another class of states is Greenberger-Horne-Zeilinger (GHZ) states [81], which are maximally entangled states across an arbitrary number of qubits, n, of the form

$$|\psi\rangle_{\text{GHZ}}^{(n)} = \frac{1}{\sqrt{2}}(|0\rangle^{\otimes n} + |1\rangle^{\otimes n}). \qquad (13.5)$$

GHZ states are useful for various quantum information processing applications, including quantum anonymous broadcasting. These states are particularly susceptible to loss, because the loss of a single qubit completely decoheres the state

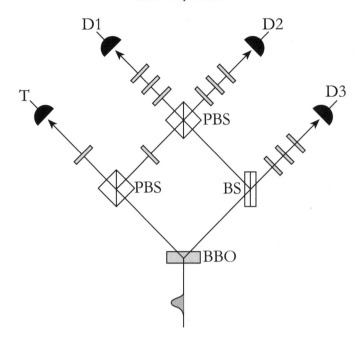

Figure 13.4 Linear optics circuit for the nondeterministic preparation of three-qubit polarisation-encoded GHZ states. A BBO SPDC source is pumped into the double-excitation regime. Following evolution through the linear optics network, measurement of a photon in the tigger mode (T) heralds the preparation of a GHZ state across modes $D1$, $D2$ and $D3$.

into a perfect mixture of the $|0\rangle^{\otimes(n-1)}$ and $|1\rangle^{\otimes(n-1)}$ states, with complete loss of entanglement and coherence,

$$
\hat{\rho}_{\text{GHZ}}^{\text{loss}} = \text{tr}_1\left(|\psi\rangle_{\text{GHZ}}^{(n)}\langle\psi|_{\text{GHZ}}^{(n)}\right)
$$

$$
= \frac{1}{2}\left(|0\rangle^{\otimes(n-1)}\langle0|^{\otimes(n-1)} + |1\rangle^{\otimes(n-1)}\langle1|^{\otimes(n-1)}\right). \tag{13.6}
$$

A simple linear optics circuit for the preparation of three-qubit polarisation-encoded GHZ states is shown in Figure 13.4.

13.5 Bell States

Bell states, also known as Einstein-Podolsky-Rosen (EPR) pairs [63], which are maximally entangled two-qubit states, are particularly useful for many applications, including quantum teleportation, cluster state preparation and entanglement swapping. Bell states are the special case of two-qubit cluster states or, equivalently, two-qubit GHZ states,

$$|\Phi^+\rangle = |\psi\rangle_{\text{GHZ}}^{(2)}. \tag{13.7}$$

Bell pairs may be directly prepared as the two-mode output from a type-II[1] SPDC source or using nondeterministic linear optics from single-photon sources.

There are four Bell states,[2] defined as

$$|\Phi^\pm\rangle = \frac{1}{\sqrt{2}}(|0\rangle_A|0\rangle_B \pm |1\rangle_A|1\rangle_B),$$

$$|\Psi^\pm\rangle = \frac{1}{\sqrt{2}}(|0\rangle_A|1\rangle_B \pm |1\rangle_A|0\rangle_B), \tag{13.8}$$

which are locally equivalent to one another via the application of Pauli operators and may therefore be transformed to one another without classical or quantum communication between the two parties. Specifically,

$$|\Phi^+\rangle = \hat{Z}|\Phi^-\rangle = \hat{X}|\Psi^+\rangle = \hat{X}\hat{Z}|\Psi^-\rangle, \tag{13.9}$$

where $\hat{X}$ and $\hat{Z}$ could apply to either qubit, up to global phase.

In Chapter 17 we present the case that these states are so useful on their own that one might be justified in building entire quantum networks based purely on the distribution of Bell pairs. This is the basis for *quantum repeater networks*, which will be discussed in Chapter 18.

13.6 Squeezed States

Of particular interest to metrology and continuous-variables quantum computing in particular are squeezed states, states that have been longitudinally distorted in phase space. In the metrological context, squeezed states enable sub-shot-noise limited metrology, thereby outperforming any classical protocol using, for example, coherent states.

Mathematically, squeezing is represented using the squeezing operator introduced in Eq. (6.21). Experimentally, such states are prepared using nonlinear crystals. It is intuitively obvious that linear optics alone cannot prepare such states, owing to the nonlinear terms in the definition of the operator, which do not preserve photon number and therefore cannot be passive.

---

[1] In type II SPDC the photon pair is polarisation-entangled, directly preparing a Bell pair in the polarisation basis. In type I SPDC both photons have the same polarisation, yielding only photon number correlation but no polarisation entanglement.

[2] $|\Psi^-\rangle$ is also referred to as a *singlet* state and $|\Psi^+\rangle$ as a *triplet* state.

Of particular interest are squeezed coherent states, $\hat{S}(\xi)|\alpha\rangle$, which are minimum uncertainty states, saturating the Heisenberg uncertainty relation. A special case of this is squeezed vacuum states, $\hat{S}(\xi)|0\rangle$, which are even-parity states (i.e., containing strictly even photon number terms). This implies that, like cat states, they are very vulnerable to decoherence for the same reason.

14

Measurement

As a last (and possibility also intermediate) step in any quantum protocol is the measurement of quantum states. State measurement is, in the most general context, essentially state preparation in reverse and brings with it many of the same challenges.

Different detection schemes bring with them their own (potentially substantial) costs and technological challenges. State-of-the-art micropillar photodetectors, at the time of writing, cost on the order of $100Ks and require a sophisticated laboratory setup. Clearly this type of infrastructure is inaccessible to many players, and borrowing or licensing access to such equipment over a quantum network would pave the way for broader accessibility to state-of-the-art technology.

Each type of state being measured, in combination with the nature of the detection scheme, brings with it its own limitations. Specifications of interest include dead time, speed (relevant when implementing feedforward) and spatiospectral filtering characteristics.

These represent significant technological challenges, which are costly to overcome, necessitating outsourcing over the quantum internet to become economically viable on a large scale. Future quantum networking protocols must accommodate in a very generic and future-proofed sense the error metrics covering all of the above error models, enabling reliable, predictable quality of service for outsourced quantum measurement.

With the ability to perform measurements over a complete basis for the respective system, quantum state tomography, and consequently quantum process tomography, can also be outsourced, because both of these protocols are built entirely on determining measurement expectation values in some known basis.

14.1 Photodetection

Perhaps the most useful, and ubiquitous, type of optical state measurement is photodetection, where we would like to count photon number. Broadly, there are two
main classes of photodetectors – *number-resolved* and *non-number-resolved* (or
'bucket' detectors). These behave exactly as the names suggest, with the former
typically being more expensive and technologically demanding than the latter.

Mathematical Representation

A completely general detector can be modelled as a positive operator-valued measure (POVM),

$$\hat{\Pi}_m = \sum_{n=0}^{\infty} P(m|n)|n\rangle\langle n|, \tag{14.1}$$

where $P(m|n)$ is the conditional probability of measuring m photons given n inci-
dent photons. The POVM is fully characterised by the conditional probabilities,
which must be inferred from the specifics of the architecture. Alternatively, a quantum process formalism can be constructed as

$$\mathcal{E}_m(\hat{\rho}) = \sum_{n=0}^{\infty} P(m|n)\hat{E}_n\hat{\rho}\hat{E}_n^{\dagger}, \tag{14.2}$$

where $\hat{E}_n = \hat{E}_n^{\dagger} = |n\rangle\langle n|$ are the Kraus operators.

   These mathematical representations very conveniently reduce the characterisation and representation of photodetectors to calculating the matrix of conditional
probabilities, $P(m|n)$. This readily allows various experimental effects and imper-
fections to be accommodated.

Experimental Issues

The key parameter of interest in a photodetector, in addition to whether or not it
is number-resolving, is its efficiency, η, or the probability that a given incident
photon will trigger the detector. For most applications, the goal is to maximise η.
As one might expect, there is a direct trade-off between η and cost, with very high-
efficiency detectors often economically out of reach for many experimentalists.
Also of interest is the 'dark count' rate – the rate at which the detector falsely
clicks in the absence of photons. However, this is often ignored as modern detectors
typically exhibit very low dark count rates.

Mathematically, the measurement operators for inefficient number-resolved detection are

$$\hat{\Pi}_n = \eta^n \sum_{m=n}^{\infty} \binom{m}{n} (1-\eta)^{m-n} |m\rangle\langle m| \qquad (14.3)$$

for measurement outcome n in the photon number basis, and for non-number-resolved detection,

$$\hat{\Pi}_0 = \sum_{m=0}^{\infty} (1-\eta)^m |m\rangle\langle m|,$$

$$\hat{\Pi}_{>0} = \hat{I} - \hat{\Pi}_0. \qquad (14.4)$$

Thus, inefficiency results in projection onto the wrong photon number, making measurement outcomes incorrect.

In addition to their operation in the photon number basis, photodetectors exhibit spatiotemporal characteristics that affect their operation in quantum information processing protocols. For example, imperfect spectral response can undermine photonic interference, affecting which-path erasure protocols, such as Bell state projection. However, in many cases this can be improved on using spectral filtering or time-gating techniques, also at the expense of experimental complexity and resource overhead.

Furthermore, photodetectors are subject to 'dead time', which renders them inactive for a finite recovery period following a detection event. This is of special importance in time bin–encoded schemes, where detectors must resolve photons over very short timescales on the order of nanoseconds. Dead time can be modelled as a time-dependent efficiency of the form

$$\eta(t) = \begin{cases} 0, & t < \tau_{dt} \\ \eta_{ss}, & t \geq \tau_{dt} \end{cases}, \qquad (14.5)$$

where t is time, τ_{dt} is the detector's dead time, and η_{ss} is the detector's steady-state efficiency (i.e., when not dead).

Photodetectors of all types are inevitably subject to 'dark counts', whereby thermal noise, either within the detector or coupled from the noisy external environment, triggers nonexistent detection events. The distribution follows exactly that of the thermal state photon number distribution. Thus, the probability of n dark counts occurring is

$$p_{dc}(n) = e^{-|\alpha|^2} \frac{|\alpha|^2}{n!}, \qquad (14.6)$$

where α is a parametrisation of the temperature of the environmental noise. Fortunately, modern detector technology is able to keep dark count rates very low, making this far less of an issue than the aforementioned ones, with loss being dominant.

Finally, all photodetection techniques are subject to some degree of 'time jitter' – noise in the detector's reported time of detection. This can be extremely important in the context of temporal mode matching, where post-selection upon detection events in an extremely narrow time window effectively enforces temporal indistinguishability.

14.2 Multiplexed Photodetection

Number-resolved detectors are the more challenging ones to experimentally realise. However, using multiplexing techniques, non-number-resolved detectors can be used to closely approximate number resolution [67, 12, 1, 156], at the expense of an (efficient) overhead in the complexity of the experiment, which comes at a cost.

Specifically, there is a direct trade-off between the confidence in photon number outcomes and experimental overhead. The idea behind this is simple. We spread out an n-photon state evenly across a large number of modes, m, and detect each one independently using a non-number-resolved photodetector. If $m \gg n$, it is unlikely that more than a single photon will reach any given detector. Thus, by summing the total number of clicks across all detectors, we closely approximate the true photon number. This multiplexing can be performed in the spatial or temporal domain, shown in Figure 14.1, and has been a widely employed technique in laboratories without access to expensive number-resolved detectors.

Mathematically, we are interested in the probability $P(n_{\text{meas}} = n_{\text{inc}})$ that the measured number of photons (n_{meas}) matches the actual number of incident photons (n_{inc}). The structure of this expression will vary enormously depending on the details of the architecture (e.g., multiport interferometer vs fibre loop). However, [156] presented a very general mathematical formalism applicable to all architectural variants. The simplest case to consider is the multiport interferometer, owing to its perfect symmetry. The probability is simply the probability that no output mode from the multiport contains multiple photons. A quick calculation yields

$$P(n_{\text{meas}} = n_{\text{inc}}) = \frac{\eta^n m!}{m^n (m-n)!},$$ (14.7)

for efficiency η, not accounting for other lesser errors such as dark counts. For perfect efficiency, $\eta = 1$, this probability always approaches unity in the limit of a large number of modes,

$$\lim_{m \to \infty} P(n_{\text{meas}} = n_{\text{inc}}) = 1.$$ (14.8)

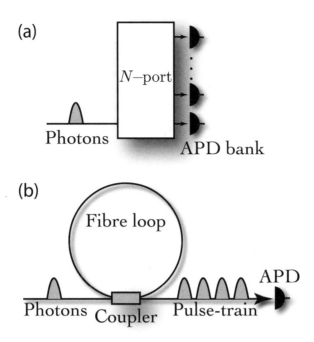

Figure 14.1 Multiplexed number-resolved photodetection using non-number-resolved photodetectors. The principle is to spread out (as uniformly as possible) a multiphoton state across a large number of modes, sufficiently large that it is unlikely that more than one photon will be present in any given mode. Then, the sum of the number of clicks at each mode closely approximates the incident photon number. (a) In the spatial domain. (b) In the temporal domain. The advantage of employing the temporally multiplexed architecture is that only a single detector is required, unlike the multiple independent detectors required in the spatially multiplexed scheme. However, this requires that the dead time of the detector is less than the round-trip time of the fibre loop. An alternate, but conceptually equivalent, approach is to spatially disperse the optical field across a charge-coupled device (CCD), much like that found in a regular digital camera, except with single-photon resolution per pixel. This achieves an effectively very large number of optical modes.

14.3 Homodyne Detection

A homodyne detector interferes a state with a coherent state on a beam splitter, which acts as a phase reference, before photodetecting both output modes and taking the difference in the photon count rates (Figure 14.2). This effectively allows us to observe 'beating' effects between the signal and reference probe.

By sweeping through the amplitude and phase of the reference beam, we are able to directly sample points in phase-space, allowing the Wigner function – which is isomorphic to the density operator – of an unknown state to be fully reconstructed.

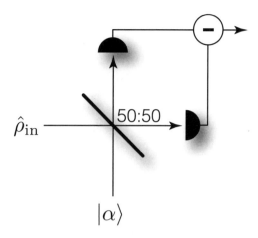

Figure 14.2 Homodyne detection of an unknown optical state $\hat{\rho}_{\text{in}}$, by mixing it with a reference coherent state $|\alpha\rangle$ on a 50:50 beam splitter and taking the difference in the photodetection rates at the output modes. By sweeping through the phase and amplitude of $|\alpha\rangle$, we can directly sample the Wigner function of $\hat{\rho}_{\text{in}}$, allowing its full reconstruction.

Equivalently, homodyne detection can directly sample the position ($\hat{x}$) and momentum ($\hat{p}$) operators or arbitrary linear combinations of the two.

The operation of homodyne measurement is most easily visualised in phase-space, where it can be regarded as integrating along an infinite line with arbitrary rotation determined by the phase reference.

This measurement technique is typically applied to continuous-variable states rather than photon number states. Though conceptually straightforward, preparing the reference beam requires a coherent source, which can become costly.

14.4 Bell State and Parity Measurements

For the purposes of which-path erasure, essential for optical cluster state preparation and quantum teleportation, Bell state measurements (i.e., projections onto the Bell basis given in Eq. (13.8)) or, equivalently, parity measurements are important.

To realise this, there are two primary options. The first is to use a controlled-NOT (CNOT) gate, which is necessarily nondeterministic using linear optics. The second is to perform a *partial* Bell state projection using a polarising beam splitter (PBS) – a beam splitter that completely transmits vertical polarisation and completely reflects horizontal polarisation [33]. In the Heisenberg picture, the transformation of the photonic creation operators implemented by a PBS is

$$\hat{h}_1^\dagger \rightarrow \hat{h}_2^\dagger,$$
$$\hat{h}_2^\dagger \rightarrow \hat{h}_1^\dagger,$$

$$\hat{v}_1^\dagger \to \hat{v}_1^\dagger,$$
$$\hat{v}_2^\dagger \to \hat{v}_2^\dagger, \tag{14.9}$$

where $\hat{h}_i^\dagger$ ($\hat{v}_i^\dagger$) are the horizontal (vertical) creation operators for the ith mode. The measurement projectors implemented by the PBS, when both modes are measured in the diagonal $(+/-)$ basis,[1] are then

$$\hat{\Pi}_{\mathrm{Bell}}^+ = |H,H\rangle\langle H,H| + |V,V\rangle\langle V,V|,$$
$$\hat{\Pi}_{\mathrm{Bell}}^- = |H,H\rangle\langle H,H| - |V,V\rangle\langle V,V|,$$
$$\hat{\Pi}_{\mathrm{HV}} = |H,V\rangle\langle H,V|,$$
$$\hat{\Pi}_{\mathrm{VH}} = |V,H\rangle\langle V,H|, \tag{14.10}$$

where the former two represent successful projection onto the Bell basis and the latter two represent failures, effectively measuring both modes in the H/V basis.

Technically, $\hat{\Pi}_{\mathrm{Bell}}^\pm$ are not Bell measurements but rather projections onto the even parity subspace. A true Bell projection would implement $|\Phi^\pm\rangle\langle\Phi^\pm|$. However, in an optical context the two terms are often used interchangeably, because they exhibit effectively the same behaviour, given that the detection process is destructive.

Bell projections using CNOT gates can be implemented with arbitrarily high success probability in principle. However, in most scenarios of interest (such as cluster state preparation and entanglement purification) Bell projection using a PBS succeeds with probability of $1/2$, because a PBS is only able to uniquely distinguish two of the four Bell states. To its advantage, such 'partial' Bell measurements only require high HOM visibility, avoiding the need for the interferometric stability inherent internally within linear optics quantum computing (LOQC) CNOT gates.

Though partial Bell state projection using a PBS is relatively straightforward, LOQC CNOT gates (which are very desirable owing to their near-determinism) are very technologically challenging, with drastic resource overheads, particularly for high success probability. Thus, outsourcing them to the cloud may be very economically efficient.

---

[1] By measuring in the diagonal basis we erase information about whether photons were horizontally or vertically polarised, thereby projecting onto the coherent subspace of both possibilities. Such a diagonal basis measurement may be implemented using a wave plate to perform a Hadamard polarisation rotation, followed by another PBS, separating the horizontal and vertical components, which are then independently measured via regular photodetection.

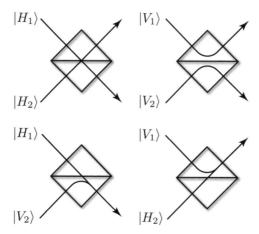

Figure 14.3 Partial Bell state projection using a polarising beam splitter (PBS). The PBS completely transmits horizontally polarised light, whilst completely reflecting vertically polarised light. Shown are the four possible two-photon input states and the respective trajectories followed by the photons. To complete the partial Bell projection we measure the output modes in the diagonal basis, $|\pm\rangle = \frac{1}{\sqrt{2}}(|H\rangle \pm |V\rangle)$, such that horizontally and vertically polarised photons cannot be distinguished. If the input state was $|H, H\rangle$ or $|V, V\rangle$, we would measure one photon in each output mode (both transmitted or both reflected). Because the detectors cannot distinguish $|H\rangle$ from $|V\rangle$, this effectively projects us onto the coherent superposition of both possibilities ('which-path erasure'), implementing the measurement projector $\hat{\Pi}_{\mathrm{Bell}}^{\pm} = |H, H\rangle\langle H, H| \pm |V, V\rangle\langle V, V|$. If, on the other hand, we measure two photons at one output mode, we know with certainty what the polarisations of both incident photons were and we probabilistically implement one of the projectors $\hat{\Pi}_{\mathrm{HV}} = |H, V\rangle\langle H, V|$ or $\hat{\Pi}_{\mathrm{VH}} = |V, H\rangle\langle V, H|$, effectively performing polarisation-resolved detection on both modes, which equates to a $\hat{Z}$ measurement on the logical qubits. The practical outcome of this is that, when using a PBS to prepare cluster states, with probability $1/2$ we are able to successfully fuse two smaller cluster states together into a larger one and with probability $1/2$ we fail to do so, instead removing two qubits from the clusters.

15

Evolution

The evolution of optical states represents an extremely broad category of quantum operations, including passive linear optics, post-selected linear optics; nonlinear optics, and light–matter interactions. Clearly, the items in this list present technological challenges, inaccessible to many users.

The error models in the evolution of optical states are largely accounted for by those discussed in Chapter 7.

15.1 Linear Optics

Linear optics networks [176] implement unitary linear maps on the photonic creation operators of the form

$$\hat{U}\hat{a}_i^\dagger\hat{U}^\dagger \rightarrow \sum_{j=1}^{m} U_{i,j}\hat{a}_j^\dagger, \tag{15.1}$$

where $\hat{a}_i^\dagger$ is the photonic creation operator on the ith of the m modes and U may be any SU(m) matrix. It was shown by [139] that arbitrary transformations of this form may be decomposed into $O(m^2)$ linear optical elements (beam splitters and phase shifters), enabling efficient construction of arbitrary linear transformations. Furthermore, the algorithm for determining the decomposition of such transformations has polynomial classical runtime (i.e., residing in **P**). Note that the original Reck *et al.* decomposition is not unique, and various other topologies of optical elements also enable universality, each with their own implementational advantages and disadvantages.

Each individual beam splitter in such a decomposition is an arbitrary SU(2) matrix acting on two photonic creation operators, $\hat{a}^\dagger$ and $\hat{b}^\dagger$,

$$\begin{pmatrix} \hat{a}_{\text{out}}^\dagger \\ \hat{b}_{\text{out}}^\dagger \end{pmatrix} = \begin{pmatrix} e^{i\phi_1}\sqrt{\eta} & e^{i\phi_2}\sqrt{1-\eta} \\ e^{-i\phi_2}\sqrt{1-\eta} & -e^{-i\phi_1}\sqrt{\eta} \end{pmatrix} \begin{pmatrix} \hat{a}_{\text{in}}^\dagger \\ \hat{b}_{\text{in}}^\dagger \end{pmatrix}, \tag{15.2}$$

where $0 \leq \eta \leq 1$ is the reflectivity and $0 \leq \phi_1, \phi_2 \leq 2\pi$ determine the phase relationships.

When operating in the polarisation basis, wave plates enable the same transformation as beam splitters do in dual-rail encoding. The phase shifters implement the unitary operation

$$\hat{\Phi}(\phi) = e^{-i\phi\hat{n}} \tag{15.3}$$

or, equivalently, in the Heisenberg picture,

$$\hat{U}\hat{a}^\dagger\hat{U}^\dagger \rightarrow e^{-i\phi}\hat{a}^\dagger, \tag{15.4}$$

for phase shift ϕ.

These linear optics evolutions are most commonly implemented using either:

• Bulk optics: discrete optical elements are arranged on an optical table.
• Time bin architectures: time bin–encoded qubits evolve through delay lines and interfere at a single central optical component.
• Integrated waveguides: all passive components are etched into a chip. When optical modes are brought physically close together, evanescent coupling allows photons to coherently hop between modes. This gives rise to evolution described by the coupled oscillator Hamiltonian, where the coupling coefficients are dictated by the proximity and geometry of the waveguides.

15.2 Nonlinear Optics

Aside from the linear transformations described above, which are passive and photon number–preserving, various active, nonlinear interactions are also of interest to optical quantum information processing. The most prominent of these are primarily considered as transformations in phase-space, using, for example, the Wigner function representation.

The most well-known nonlinear transformation is the displacement operation, which translates the Wigner function by some arbitrary amplitude in phase-space, whilst preserving all other features of the phase-space representation. This is described by the unitary operator

$$\hat{D}(\alpha) = \exp\left[\alpha\hat{a}^\dagger - \alpha^*\hat{a}\right], \tag{15.5}$$

where α is the displacement amplitude. This transformation is easily implemented by mixing a state on a low-reflectivity beam splitter with a coherent state of some arbitrary complex amplitude, which determines the displacement amplitude. In the special case of a displacement operator acting on the vacuum state, we obtain a coherent state of equal amplitude, $\hat{D}(\alpha)|0\rangle = |\alpha\rangle$.

Another common nonlinear transformation is squeezing, discussed in Section 13.6. This implements the unitary operator

$$\hat{S}(\xi) = \exp\left[\frac{1}{2}(\xi^*\hat{a}^2 - \xi\hat{a}^{\dagger 2})\right],$$ (15.6)

where ξ is the squeezing parameter, which has the effect of applying a dilation of some arbitrary factor along a particular axis in phase-space.

Thus, jointly, the displacement and squeezing operators enable arbitrary translations and dilations in phase-space. These operations form the basis for continuous-variable quantum computing schemes.

16

High-Level Protocols

Building upon the aforementioned primitive protocols for quantum networking, we can construct a plethora of higher-level protocols that implement more powerful end-user applications. These high-level protocols are ubiquitous in quantum information processing and form building blocks for even more powerful architectures, such as full cloud quantum computing, to be discussed in Chapter 29.

16.1 Random Number Generation

"God does not play dice!" — Albert Einstein.

Perhaps the simplest quantum information processing task is that of perfect random number generation. True random numbers have widespread applications in cryptography, Monte Carlo simulations and any type of randomised (e.g., **BPP**) algorithm.

Classical random number generators are actually deterministic, following the laws of classical physics, but so difficult to predict that we accept them to be as good as random. But for some applications this is not enough, and we must make sure that no correlations of any type exist between different random numbers or between the random numbers and their environment.

This can be achieved in many different ways quantum mechanically. Ultimately, they are all based on the Heisenberg uncertainty principle, that certain quantum mechanical measurements yield uncertainty. A simple optical implementation is shown in Figure 16.1.

The cynics amongst us might question the nondeterminism of the laws of Nature and ask whether quantum random numbers really are truly random (in the sense of nondeterminism) or whether they also are just too hard to predict that we treat them as effectively random. The answer to this is that it has been proven that quantum mechanics is inconsistent with 'hidden variable theories'; i.e., that there

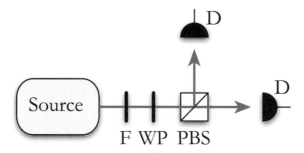

Figure 16.1 A simple optical implementation for a quantum binary random number generator. The source prepares single photons, which pass through a polarisation filter (*F*) to retain only horizontally polarised photons, $|H\rangle$. A wave plate (*WP*) then rotates the polarisation into the diagonal basis, yielding a stream of $|+\rangle$ polarised photons. These are then separated into horizontal and vertical components via a polarising beam splitter (*PBS*), each of which is independently photodetected (*D*). If correctly implemented, the two detectors click one at a time, each with 50% probability, and the random stream is inherently nondeterministic, guaranteed by the nondeterminism of quantum mechanics. Random numbers are produced at a rate of one bit per photodetection.

is an underlying, but inaccessible, determinism in the world that is guiding quantum measurements in a completely deterministic manner. This disproof effectively validates the notion of quantum mechanical perfect random number generation.

Consider the scenario where a client needs a stream of true random numbers for use in her Monte Carlo simulation algorithm or as a secret key for her email encryption. She has limited quantum resources herself, so she outsources it to her better-equipped mate. Depending on her own resource limitations and potential security considerations, her friend could either (1) implement the full protocol described above, providing her with a classical random bitstream, or (2) only take care of photon generation, providing her with a perpetual source of high-quality photons for her to measure herself using a simple photodetector. (1) and (2) would both be suitable if the intention was to apply the source of randomness to a Monte Carlo simulation. But in a cryptographic scenario, where the randomness is being used for key generation, clearly Alice could not outsource the measurement stage without revealing her secret key. In this instance, Bob can act as the provider of photons, and Alice does the measurements herself to keep her random bit string secret.

For cryptographic purposes, there are far more stringent constraints placed upon our random number generator than for use in, say, a Monte Carlo computer simulation – cryptographic random number generation. In this context, the demands placed on the amount of bias or correlations in the random number stream are very stringent. An enormous amount of research has been invested into this topic, and sophisticated statistical tests have been developed for establishing crypto-worthiness of a random number stream. The fact that quantum random numbers, via the inherent nondeterminism of quantum mechanics, do not obey any hidden variable theory implies that there is intrinsically no underlying 'seed' to the random number stream that reveals the entire deterministic sequence. This is unlike any classical generator, where there always is such a seed but it is simply taken to be too hard to determine.

This scenario is an obvious example of where a User Datagram Protocol (UDP)-like SEND-AND-FORGET protocol may be viable. Unlike most other applications, Bob is broadcasting a stream of identical, pure quantum states that are not entangled with any peripheral system and are easily replicated, with no correlations between distinct photons. Thus, if any particular photon fails to reach Alice, it matters not, because she can simply await the next one emanating from Bob's bombardment of photons (the 'shotgun' approach). There are no quality of service (QoS) requirements.

16.2 Entanglement Purification

Entangled states, most notably Bell pairs, play a central role in many quantum technologies. These maximally entangled states are easily represented optically using polarisation encoding of single photons and can be nondeterministically prepared directly using SPDC, or post-selected linear optics.

Bell pairs are the basis for building cluster states, some quantum cryptography protocols and quantum teleportation, to name just a few applications. Therefore, distributing entangled states with the highest entanglement metrics is extremely important. In short, entanglement can be considered a valuable quantum resource upon which many other protocols may be built.

Suppose that Alice and Bob share an entangled pair. Quantum mechanics, specifically the very definition of entanglement itself, prohibits local operations performed by Alice and Bob from increasing the level of entanglement. However, if Alice and Bob share multiple pairs, they can perform an operation known as *entanglement purification* or *entanglement distillation*, whereby two lower-fidelity entangled pairs are consumed and projected onto a single entangled pair with higher fidelity [19, 25, 51]. Such protocols will be extremely useful in protocols where achieving the highest possible degree of entanglement is paramount; for example, when error thresholds must be achieved for the purpose of error correction and fault tolerance [127].

Taking two polarisation-encoded photonic Bell pairs, say, $|\Psi^+\rangle$, and subjecting them to a dephasing error model yields a mixed state of the form

$$\hat{\rho}_{\text{in}} = F|\Psi^+\rangle\langle\Psi^+| + (1 - F)|\Psi^-\rangle\langle\Psi^-|, \tag{16.1}$$

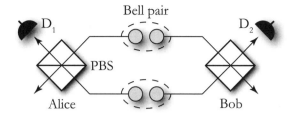

Figure 16.2 Elementary entanglement purification using linear optics. Two Bell pairs are distributed between Alice and Bob, each of which has been subject to a dephasing error model. Alice and Bob perform Bell measurements on their two qubits using a PBS and polarisation-resolved photodetection (D_1 and D_2). Upon successful Bell state projection (Bell measurements are necessarily nondeterministic using linear optics), Alice and Bob will share a single Bell pair with higher fidelity than the two input pairs.

where F is the entanglement fidelity, which is a function of the dephasing rate. Note that $|\Psi^+\rangle$ and $|\Psi^-\rangle$ are related by local Pauli phase-flip operations ($\hat{Z}$) applied to either qubit,

$$|\Psi^-\rangle = \hat{Z}_A|\Psi^+\rangle = \hat{Z}_B|\Psi^+\rangle. \tag{16.2}$$

A linear optics entanglement purification protocol can be simply implemented using two polarising beam splitters (PBSs) [131, 130]. Alice uses one PBS to interfere the photons from her side of each of the photon pairs, measuring one output only, which implements a nondeterministic, partial Bell state projection. Bob does the same on his side. What is left is one photon in Alice's hands and one in Bob's. When successful, they will now be sharing a single entangled pair of higher Bell state fidelity than the two starting states. The protocol is shown in Figure 16.2.

Note that when using PBSs to perform the Bell projections, the protocol is necessarily nondeterministic, because PBSs are only able to distinguish two of the four Bell states. Thus, each PBS has a success probability of $1/2$. And there are two PBSs per instance of the protocol; therefore, the net success probability is $1/4$. When concatenated, n applications of the protocol thus have an exponentially low success probability of $1/4^n$. This could be overcome using deterministic controlled-NOT (CNOT) gates, but these are challenging using linear optics.

Furthermore, the protocol consumes two Bell pairs upon each trial, only one quarter of which are successful. Thus, on average, eight Bell pairs are consumed for every purified Bell pair prepared, and the expected number of Bell pairs required to perform n iterations of entanglement purification grows exponentially as 8^n.

Specifically, the relationship between the input (F_{in}) and output (F_{out}) fidelities of the protocol is

$$F_{\text{out}} = \frac{F_{\text{in}}^2}{F_{\text{in}}^2 + (1 - F_{\text{in}})^2}. \tag{16.3}$$

Note that there is a breakeven point above which the protocol strictly increases fidelity and below which strictly decreases it. This occurs at $F_{\text{in}} = 1/2$. Provided that pairs can be communicated above this fidelity threshold, bootstrapped application of the protocol could be employed to boost entanglement fidelity asymptotically close to unity (but with exponential resource overhead, because each operation nondeterministically consumes two pairs to produce one). But below this threshold it is impossible to recover any more entanglement than we started with. This provides an example of an application where the protocol being implemented dictates strict requirements on network cost metrics. Specifically, assuming perfect Bell pairs to begin with, the routes by which they are communicated must strictly ensure entanglement fidelities of at least $F = 1/2$ upon reaching their destination. Here, a type of ALL OR NOTHING networking strategy would be applicable – if the fidelity requirement is not met, the state cannot be purified and might as well be thrown away to make way for other traffic.

A theoretical analysis of this protocol has been performed, accounting for mode-mismatch in the protocol [153], where it was found that mode mismatch shifts the breakeven point upwards and lowers the maximum value of F_{out} – with more mode mismatch, a higher starting fidelity is required to break even and we achieve a lower, sub-unity output fidelity. In this case, a cost function that combines the dephasing and mode mismatch metrics of the network will be required.

Importantly, this protocol is based on partial Bell state measurement and therefore does not require interferometric stability, only high HOM visibility, thus making stabilisation comparatively easy over long distances.

Entanglement purification can also be performed using physical encodings other than single photons. For example, this has been demonstrated using Gaussian continuous-variable quantum states [57].

16.3 Quantum State Teleportation

Quantum state teleportation [22] is an essential ingredient in many higher-level protocols. It forms the basis of cluster state quantum computing, some quantum error correction (QEC) codes and linear optics quantum computing and can act as a mediator for long-range transmission of quantum states, amongst others.

In the standard teleportation protocol, Alice begins with a single qubit,

$$|\phi\rangle = \alpha|0\rangle + \beta|1\rangle, \tag{16.4}$$

which she would like to teleport to Bob. Importantly, no quantum communication between the two is allowed, because obviously this would make the problem trivial.

However, classical communication is allowed (and turns out to be necessary) and, furthermore, they share an entangled Bell pair as a resource. Thus, Alice begins with two qubits and Bob begins with one – his half of the entangled pair onto which Alice's state ought to be teleported. The initial state is therefore

$$
|\psi\rangle_{\text{in}} = |\phi\rangle_{A_1}|\Psi^+\rangle_{A_2,B}
$$

$$
= \frac{1}{\sqrt{2}}(\alpha|0\rangle_{A_1} + \beta|1\rangle_{A_1})(|0\rangle_{A_2}|1\rangle_B + |1\rangle_{A_2}|0\rangle_B). \tag{16.5}
$$

The first step of the protocol is for Alice to perform a two-qubit entangling measurement on her two qubits, projecting onto the Bell basis, Eq. (13.8). She obtains one of four measurement outcomes. For illustration, suppose that she measures the $|\Psi^+\rangle$ outcome. Then the projected state is

$$
|\psi\rangle_{\text{proj}}^{\Psi^+} = \langle\Psi^+|_{A_1,A_2}|\psi\rangle_{\text{in}}
$$

$$
= \frac{1}{\sqrt{2}}\langle\Psi^+|_{A_1,A_2}|\psi\rangle_{A_1}(|0\rangle_{A_2}|1\rangle_B + |1\rangle_{A_2}|0\rangle_B)
$$

$$
= \frac{1}{2}\big(\langle0|_{A_1}\langle1|_{A_2} + \langle1|_{A_1}\langle0|_{A_2}\big)\cdot
$$

$$
\big(\alpha|0\rangle_{A_1} + \beta|1\rangle_{A_1})(|0\rangle_{A_2}|1\rangle_B + |1\rangle_{A_2}|0\rangle_B\big)
$$

$$
= \frac{1}{2}(\alpha|0\rangle_B + \beta|1\rangle_B)
$$

$$
= \frac{1}{2}|\phi\rangle_B, \tag{16.6}
$$

which is Alice's initial state. For all four possible Bell measurement outcomes we have

$$
|\psi\rangle_{\text{proj}}^{\Psi^+} = \frac{1}{2}(\alpha|0\rangle_B + \beta|1\rangle_B)
$$

$$
= \frac{1}{2}|\phi\rangle_B,
$$

$$
|\psi\rangle_{\text{proj}}^{\Psi^-} = \frac{1}{2}(\alpha|0\rangle_B - \beta|1\rangle_B)
$$

$$
= \frac{1}{2}\hat{Z}|\phi\rangle_B,
$$

$$
|\psi\rangle_{\text{proj}}^{\Phi^+} = \frac{1}{2}(\alpha|1\rangle_B + \beta|0\rangle_B)
$$

$$
= \frac{1}{2}\hat{X}|\phi\rangle_B,
$$

$$|\psi\rangle_{\text{proj}}^{\Phi^-} = \frac{1}{2}(\alpha|1\rangle_B - \beta|0\rangle_B)$$

$$= \frac{1}{2}\hat{X}\hat{Z}|\phi\rangle_B, \tag{16.7}$$

which are all locally equivalent to $|\phi\rangle$ under Pauli gates and can be corrected by Bob, given communication of the classical Bell measurement outcome provided by Alice. The full protocol is described in Box 16.1.

Box 16.1 **Quantum state teleportation of a single qubit.**

function StateTeleportation($|\phi\rangle_{A_1}$, $|\Phi^+\rangle_{A_2,B}$) :

1. Alice prepares the state $|\phi\rangle_{A_1}$, which she would like to teleport to Bob.
2. Alice and Bob share the Bell pair $|\Phi^+\rangle_{A_2,B}$.
3. Alice performs a Bell state projection between qubits A_1 and A_2.
4. Alice communicates the classical measurement outcome to Bob - one of four outcomes.
5. Bob applies an appropriate local correction to his qubit - some combination of the Pauli operators $\hat{X}$ and $\hat{Z}$ - according to the classical measurement outcome

$$|\Psi^+\rangle\langle\Psi^+| \rightarrow \hat{I},$$
$$|\Psi^-\rangle\langle\Psi^-| \rightarrow \hat{Z},$$
$$|\Phi^+\rangle\langle\Phi^+| \rightarrow \hat{X},$$
$$|\Phi^-\rangle\langle\Phi^-| \rightarrow \hat{Z}\hat{X}. \tag{16.8}$$

6. Bob is left with the state $|\phi\rangle_B$.

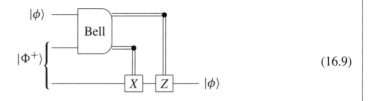

$$(16.9)$$

In general, the protocol is deterministic, although using PBSs to perform partial Bell measurements the success probability is at most $1/2$.

The question now is what error metrics apply and how they accumulate in the teleportation protocol. The answer is straightforward – the final teleported qubit accumulates all local Pauli errors (e.g., dephasing or depolarisation) associated with Alice's input state as well as any that acted upon the shared Bell pair. That is, the errors get teleported along with the state being teleported, plus any errors on the Bell pair.

In the case of loss, loss of either of Alice's qubits will immediately be detected when she performs her Bell measurement. Thus, loss becomes a located error, and the knowledge of the error allows the associated packet to be discarded and the sender and recipient are notified. On the other hand, loss of Bob's qubit will behave no differently than loss acting on an ordinary qubit channel.

Thus, in terms of Pauli errors, no special treatment is required by the network protocol – it is almost as if the teleportation protocol were not there. And in terms of loss, the Bell state projection diagnoses lost qubits, allowing appropriate action to be taken, which is actually better than if the error were undiagnosed. These are often referred as *located* and *unlocated* errors.

The total resources required to teleport a single-qubit state are the following:

1. The qubit to be teleported.
2. A shared Bell pair.
3. A two-qubit entangling measurement in the Bell basis.
4. The transmission of two classical bits.
5. Two classically controlled Pauli gates for correction.

This is more costly than sending the qubit directly over a quantum channel but may be the only approach if a direct link is not available. In the context of an internet where entanglement distribution is treated as the fundamental resource, state teleportation is the natural approach for communicating quantum states, because no quantum communication of any kind is required once the two parties have a shared Bell pair between them.

The important feature of this protocol to note is that there is no direct quantum communication between Alice and Bob, only a classical communications channel. Rather, the Bell pair mediates the transfer of quantum information, despite there being no direct quantum channel between Alice and Bob.

Relying on teleportation rather than direct quantum communication makes frugal use of quantum channels, because there is no need for direct quantum routes between every pair of nodes in the network. Instead, each node need only have a direct one-way quantum link with the central authority responsible for entanglement distribution, thereby significantly reducing the complexity of the topology of the quantum network.

The Bell state measurement can be implemented either using a CNOT gate or as a nondeterministic partial Bell state measurement using a PBS, both of which are nondeterministic using purely linear optics.

The above describes quantum state teleportation at the level of single qubits. However, when dealing with more general quantum data, which may have multi-qubit payloads, we may wish to teleport an entire 'packet' (some arbitrary quantum state). This is implemented as a simple extension of the above procedure – we simply implement n multiple independent teleportation protocols to all of the packet's n constituent qubits. Via linearity, although the teleportation protocols are being applied independently to each qubit, the net packet teleportation operation will preserve their joint state, including entanglement between them. Note, however, that if the qubit state teleportation protocols are individually nondeterministic with success probability p_{teleport}, the net success probability for the teleportation of the entire packet scales inverse exponentially with n, as $p_{\text{teleport}}{}^{n}$.

Open Destinations

In the standard quantum state teleportation protocol there is a single sender and a single receiver. A generalisation of the protocol is *open destination quantum state teleportation*, whereby there is still just one sender but any number of potential recipients. At the time of transmission, the sender does not specify the recipient but rather wishes to 'broadcast' the state to *all* recipients, such that any *one* of them can subsequently read out the state. Note that only a single recipient may actually perform the readout, because multiple readouts would violate the no-cloning theorem. However, the key new feature introduced by this variant of the protocol is that the final choice of which recipient performs the readout need not be known in advance but can be decided at an arbitrary later stage, well after the sender has completed their side of the protocol.

To implement this protocol, outlined in Figure 16.3, rather than first distributing a Bell pair between sender and recipient, we distribute an n-party GHZ state between the sender and the $n - 1$ recipients,

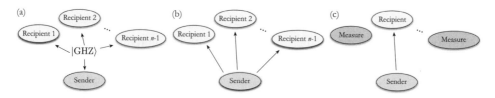

Figure 16.3 Open-destination quantum state teleportation. (a) GHZ state distribution from a central server between the sender and all potential recipients. (b) Sender performs quantum state teleportation, resulting in the state being teleported to all recipients in redundantly-encoded form. (c) All non-receivers measure out their qubits in the $\hat{X}$-basis, resulting in the teleported state arriving at the destination of just the chosen true receiver.

$$|\psi_{\text{GHZ}}^{(n)}\rangle = \frac{1}{\sqrt{2}}(|0\rangle^{\otimes n} + |1\rangle^{\otimes n}). \tag{16.10}$$

Next, note that performing an $\hat{X}$-basis measurement on one qubit in a GHZ state reduces it to an $n - 1$-qubit GHZ state, up to a potential local $\hat{Z}$-correction,

$$\langle +|\psi_{\text{GHZ}}^{(n)}\rangle = |\psi_{\text{GHZ}}^{(n-1)}\rangle,$$
$$\langle -|\psi_{\text{GHZ}}^{(n)}\rangle = \hat{Z}|\psi_{\text{GHZ}}^{(n-1)}\rangle. \tag{16.11}$$

Performing this contraction repeatedly ultimately reduces us all the way down to a single Bell pair, because

$$|\psi_{\text{GHZ}}^{(2)}\rangle = |\Phi^+\rangle. \tag{16.12}$$

Thus, to teleport from sender to recipient, all nonrecipients simply measure their qubits in the $\hat{X}$-basis and publicly report their classical measurement outcomes for the purposes of performing local corrections. What is left is a Bell pair between sender and recipient. This Bell pair may then be employed in the conventional teleportation protocol to perform the teleportation.

The key observation now is that the dynamics of this entire system must be invariant under the time ordering of the $\hat{X}$-basis measurements. Thus, whether they are performed at the beginning of the protocol (thereby directly reducing us to standard two-party teleportation) or deferred until later makes no difference to the final state obtained by the recipient. This is the basis for the ability of the protocol to broadcast the teleported state, without first specifying the intended recipient.

16.4 Quantum Gate Teleportation

Using quantum *state* teleportation as a primitive building block, quantum *gate* teleportation may be implemented [80]. Here, rather than teleporting a quantum state from one physical system to another, we teleport the action of a quantum gate onto a physical system (archetypically a maximally entangling two-qubit gate, such as a CNOT gate).

The general outline of the derivation of the protocol for teleporting a CNOT gate onto a two-qubit state is shown in Box 16.2.

Box 16.2 Teleporting a CNOT gate onto a two-qubit state.

```
function GateTeleportation (|ψ⟩_A|φ⟩_B):
```

1. We wish to apply a CNOT gate to $|\psi\rangle_A|\phi\rangle_B$.
2. Introduce two additional qubits, C and D.
3. Teleport states $|\psi\rangle_A \rightarrow |\psi\rangle_C$, $|\psi\rangle_B \rightarrow |\psi\rangle_D$.

4. Apply $\mathrm{C\hat{N}OT}|\psi\rangle_C|\phi\rangle_D$.

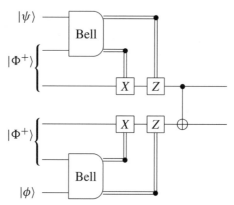

5. The CNOT is a Clifford gate and can therefore be commuted to the front of the Pauli operators to yield a CNOT followed by some different configuration of Pauli operators.

6. The CNOT now acts jointly upon the Bell pairs that were acting as a resource for the state teleportation, independent of $|\psi\rangle_A|\phi\rangle_B$.

7. Group the CNOT gate and Bell pairs together and treat them as a four-qubit resource state preparation stage, which does not depend on $|\psi\rangle_A|\phi\rangle_B$.

8. Prepare the 4-qubit resource state,
   $|\chi\rangle = \mathrm{C\hat{N}OT}_{2,3}|\Psi^+\rangle_{1,2}|\Psi^+\rangle_{3,4}$, offline in advance.

9. If the CNOT is nondeterministic, employ Repeat Until Success to prepare $|\chi\rangle$.

10. The output state is $\mathrm{C\hat{N}OT}_{C,D}|\psi\rangle_C|\phi\rangle_D$.

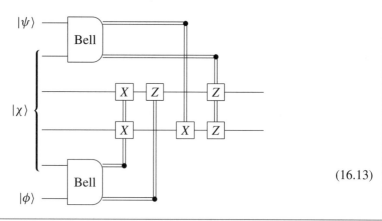

$$(16.13)$$

Most notably, gate teleportation is useful when attempting to apply two-qubit entangling operations using nondeterministic gates, in which case gate teleportation allows the nondeterministic elements to be performed offline as a resource state preparation stage, overcoming the nondeterminism during the gate application stage.

Specifically, when a CNOT gate acting directly upon two qubits fails, it corrupts those qubits, whereas if it fails during a state preparation stage, it can simply be reattempted until a success occurs, without corrupting the target qubits. A concatenated version of the gate teleportation protocol forms the basis for constructing near-deterministic entangling gates in linear optics.

Quantum gate teleportation effectively reduces the problem of implementing CNOT gates to:

1. Offline preparation of highly entangled four-qubit resource states. This need not be deterministic, because the resource state does not depend on the state to which the CNOT gate ought to be applied.
2. Two Bell measurements.
3. Some configuration of local Pauli operators, dependent on the Bell measurement outcomes.

Importantly, like quantum state teleportation, there is no need for a quantum communications channel between the two parties holding the qubits to which the gate is applied – classical communication is sufficient.

The gate teleportation idea is conceptually interesting as it converts the problem of 'gate application' to that of 'state preparation'[1] by commuting all of the entangling operations to the beginning of the protocol. Cluster state quantum computing is actually the extremity of this logic, whereby an entire quantum computation is transformed into a sequence of state and gate teleportations. One may interpret this to mean that teleportation is a universal resource for quantum computation [80].

The resource states required for gate teleportation are highly entangled four-qubit states, which are challenging to prepare, especially in the optical context. Thus, as with cluster states, if the preparation of these resource states were to be outsourced to a specialised provider, they could be in high demand.

Note that this technique works for the CNOT gate because it is a Clifford gate (i.e., it commutes with the classically controlled Pauli gates to yield a different combination of classically controlled Pauli gates). Thus, this technique does not automatically apply to *any* two-qubit gate.

---

[1] The resource state is prepared from two Bell pairs and a single CNOT gate, which is locally equivalent to a four-qubit GHZ state. In the absence of a direct source of Bell pairs, they can be prepared using separable single-qubit states and a CNOT gate. Thus, the full resource state may be prepared from separable single qubits via three CNOT gates.

16.5 Entanglement Swapping

The obvious approach to sending a qubit from Alice to Bob is to send a qubit from Alice to Bob (!). However, over long distances this may accrue impractical error rates, particularly losses. The other alternative is to employ the quantum state teleportation protocol to teleport the state between the two parties. However, this requires that Alice and Bob first share an entangled Bell pair, which must itself be distributed across the same distances. Entanglement swapping [105] is the process of taking two Bell pairs, one held by each party, and swapping the entanglement between them such that the two parties share an entangled state. This procedure can be bootstrapped to progressively swap the entanglement over longer and longer distances, yielding *quantum repeater networks*. The procedure for this protocol is shown in Box 16.3.

Box 16.3 Entanglement swapping protocol between two parties. Two Bell pairs held locally by two users, $|\Phi^+\rangle_{A_1, A_2}|\Phi^+\rangle_{B_1, B_2}$, are converted to a single Bell pair shared between the users, $|\Phi^+\rangle_{A_2, B_2}$.

```
function EntanglementSwapping (|Φ+⟩⊗2) :
```

1. Alice locally prepares the Bell pair

$$|\Phi^+\rangle_{A_1, A_2}. \tag{16.14}$$

2. Bob locally prepares the Bell pair

$$|\Phi^+\rangle_{B_1, B_2}. \tag{16.15}$$

3. The net initial state is

$$|\psi\rangle_{\text{in}} = |\Phi^+\rangle_{A_1, A_2}|\Phi^+\rangle_{B_1, B_2}. \tag{16.16}$$

4. Alice sends qubit A_1 to third-party Eve.
5. Bob sends qubit B_1 to third-party Eve.
6. Eve performs a Bell projection between A_1 and B_1, yielding

$$\langle\Phi^+|_{A_1, B_1}|\psi\rangle_{\text{in}} = |\Phi^+\rangle_{A_2, B_2}. \tag{16.17}$$

7. In the case of the other Bell projection outcomes $(\langle\Phi^-|_{A_1, B_1}, \langle\Psi^+|_{A_1, B_1} \text{ or } \langle\Psi^-|_{A_1, B_1})$, local corrections (Pauli operators) are made by Alice and/or Bob, as dictated by classical communication from Eve,

$$\langle\Phi^+|_{A_1, B_1}|\psi\rangle_{\text{in}} = |\Phi^+\rangle_{A_2, B_2},$$

$$\langle\Phi^-|_{A_1, B_1}|\psi\rangle_{\text{in}} = \hat{Z}_{B_2}|\Phi^+\rangle_{A_2, B_2},$$

$$\langle\Psi^+|_{A_1,B_1}|\psi\rangle_{\text{in}} = \hat{X}_{B_2}|\Phi^+\rangle_{A_2,B_2},$$

$$\langle\Psi^-|_{A_1,B_1}|\psi\rangle_{\text{in}} = \hat{X}_{B_2}\hat{Z}_{B_2}|\Phi^+\rangle_{A_2,B_2}. \qquad (16.18)$$

8. Alice and Bob now possess a joint Bell pair between qubits A_2 and B_2,

$$|\psi\rangle_{\text{out}} = |\Phi^+\rangle_{A_2,B_2}. \qquad (16.19)$$

$$(16.20)$$

In a sense, entanglement swapping can be regarded as 'indirect' entanglement distribution, whereby entanglement is created between two distant parties who do not directly exchange any quantum information.

Alternatively, note that the entanglement swapping is structurally almost identical to two instances of quantum state teleportation side by side. This is not a coincidence, and entanglement swapping can indeed be thought of as Bell pair state teleportation.

Now if instead of Alice and Bob we have a long chain of these operations in series, then the entanglement can be swapped across the entire length of the chain, enabling the preparation of end-to-end entangled pairs, which can be employed for state teleportation.

The advantage to this approach is that the range of each repeater can be much smaller than the entire length of the channel, easing constraints imposed by errors, notably loss. Furthermore, the entanglement swapping need not be actually performed in any chronologically linear sequence. The operations could be arbitrarily ordered, because the measurements are independent and commute. Thus, if some segments are detected as failing (e.g., qubits are lost), just those segments can be performed again without requiring the entire protocol to start from scratch, unlike the naïve direct communication technique. This DIVIDE AND CONQUER approach can drastically improve performance of the network in terms of channel capacity, improving the exponential dependence of loss on distance.

The protocol is conceptually very similar to teleportation, where instead of teleporting a qubit state we are teleporting entanglement. Because of this similarity, it inherits error propagation characteristics similar to those for teleportation discussed previously. That is, errors acting on the qubits upon which the Bell measurements are performed are effectively teleported onto the remaining qubits. Then, entanglement purification can be implemented as a higher-level layer on top of the repeaters, enabling high-fidelity entanglement distribution.

Each Bell measurement can be implemented nondeterministically using a PBS, mitigating the need for interferometric stability, as before, but therefore introducing nondeterminism into the protocol.

16.6 Superdense Coding

Superdense coding is a hybrid quantum/classical communications protocol for increasing classical bit rates between two parties who share entanglement as a resource.

Suppose that Alice wishes to send classical information to Bob over a quantum channel. The HSW theorem [87, 164] tells us that Alice can send information to Bob at a maximum rate of one bit per qubit. However, if Alice and Bob share Bell pairs, superdense coding allows information to be transmitted at a maximum rate of two bits per qubit.

Let Alice and Bob begin with the shared Bell state,

$$|B_{00}\rangle = \frac{1}{\sqrt{2}} (|0\rangle_A |0\rangle_B + |1\rangle_A |1\rangle_B), \qquad (16.21)$$

where the first qubit, A, belongs to Alice and the second qubit, B, belongs to Bob. This entangled pair is provided to them by a third-party entanglement server. The protocol exploits the fact that all four Bell states are locally equivalent and can be transformed into one another using operations performed only by Alice. Specifically, the four Bell states can be prepared from $|B_{00}\rangle$ via the local operations

$$|B_{00}\rangle = (\hat{I} \otimes \hat{I})|B_{00}\rangle$$
$$= \frac{1}{\sqrt{2}} (|0\rangle|0\rangle + |1\rangle|1\rangle),$$
$$|B_{01}\rangle = (\hat{Z} \otimes \hat{I})|B_{00}\rangle$$
$$= \frac{1}{\sqrt{2}} (|0\rangle|0\rangle - |1\rangle|1\rangle),$$
$$|B_{10}\rangle = (\hat{X} \otimes \hat{I})|B_{00}\rangle$$
$$= \frac{1}{\sqrt{2}} (|0\rangle|1\rangle + |1\rangle|0\rangle),$$
$$|B_{11}\rangle = (\hat{Z}\hat{X} \otimes \hat{I})|B_{00}\rangle$$
$$= \frac{1}{\sqrt{2}} (|0\rangle|1\rangle - |1\rangle|0\rangle). \qquad (16.22)$$

Suppose that Alice wishes to send Bob the two-bit string $x \in \{0,1\}^2$. She applies local operations on her qubit to transform the shared Bell state into the Bell state $|B_x\rangle$. There are four such states; therefore, this encodes two classical bits of information. She then sends her qubit to Bob, who already holds the other half of the entangled pair. Now by measuring in the Bell basis Bob can determine which two-bit string Alice encoded.

Note that the protocol in a sense 'cheats', because it assumes a resource of Bell pairs between Alice and Bob, which does not come for free. However, in an environment where both parties have access to the same entanglement server or repeater network, in addition to their own direct line of quantum communication, they can utilise this protocol to double classical communication rates from one bit per qubit to two.

However, this doubling in communication rate requires using quantum infrastructure, which, at least for the foreseeable future, will come at a greater cost than our present-day commodified classical hardware. It may therefore be the case that the technological effort of implementing this protocol outweighs the gain or that for the same effort other classical bandwidth-increasing technologies could be employed.

Alternatively, in a future quantum world where such technologies are cheap off-the-shelf commodities, as with our current classical ones, why not double our classical network bandwidths if we can?

16.7 Quantum Metrology

The goal of quantum metrology is to estimate an unknown phase with the greatest degree of precision. This finds many applications, perhaps most notably the recent gravity wave measurement protocols. The shot-noise limit (SNL) represents the maximum achievable precision using classical states, whereas the Heisenberg limit (HL) is the best that can be achieved using quantum resources. The goal of quantum metrology is to beat the SNL, ideally saturating the HL.

Achieving the SNL is easily done using a Mach-Zehnder interferometer fed with coherent states, which are not true quantum states. Referring to Figure 12.2, if a coherent state is inputted into one arm of the interferometer, with no phase shift ($\tau = 0$) all of the coherent amplitude would exit the corresponding output port. If, on the other hand, there were a π phase shift, all of the amplitude would exit the other output port. For intermediate τ there will be varying degrees of coherent amplitude distributed between the two outputs. Thus, the relative amplitude exiting the two output ports acts as a signature for the internal phase shift τ.

Improving upon this, HL metrology can be achieved using NOON states [56]. An alternate recent proposal (known as the MORDOR protocol, after the authors), employs only single-photon states and passive linear optics, which, although not saturating the HL, significantly beats the SNL [116].

NOON states in particular are difficult to prepare, because they cannot be deterministically prepared using linear optics, and no current source natively prepares them directly. Thus, outsourcing these state preparation stages could be of great value to end-users of metrology, were there a specialised server dedicated to this task.

16.8 Quantum-Enabled Telescopy

For the direct imaging of an object, diffraction limits the resolution of the image. When we consider two neighbouring points on the object separated by a small angle, the minimum angular separation resolvable is

$$\theta_{\min} = 1.22 \frac{\lambda}{D}, \tag{16.23}$$

known as the Rayleigh criterion, where D is the diameter of the aperture.

Current optical interferometers have limited baseline lengths and thus limited resolution. In principle one can build a telescope array with a synthetic aperture of arbitrary D; however, phase-locking the entire system is extremely difficult over long distances. A quantum information protocol has been developed to sidestep this problem.

The light arriving from the distant object is a thermal state, but the average photon number per mode is much less than 1; therefore, higher order terms are negligible. The state that reaches the telescope is therefore approximated by

$$|\psi_{\text{image}}\rangle = \frac{1}{\sqrt{2}}(|0\rangle_A|1\rangle_B + e^{i\phi}|1\rangle_A|0\rangle_B), \tag{16.24}$$

where ϕ is the relative phase shift between the two telescopes, which depends on the difference in distance of propagation. If ϕ can be measured accurately, this can give a precise estimate on the location of the object,

$$\phi = \frac{b\sin(\theta)}{\lambda}, \tag{16.25}$$

where λ is wavelength.

Often the light that arrives will be formed by a mixture of photons from different sources that emit incoherently, and different locations give rise to different phase shifts ϕ, resulting in a density matrix of the form

$$\hat{\rho} = \frac{1}{2} \begin{pmatrix} 0 & 0 & 0 & 0 \\ 0 & 1 & \mathcal{V}^* & 0 \\ 0 & \mathcal{V} & 1 & 0 \\ 0 & 0 & 0 & 0 \end{pmatrix}, \tag{16.26}$$

where $\mathcal{V}$ is the visibility, reflecting decoherence.

If we interfere the two modes at a 50:50 beam splitter, the photon will exit port 1 with probability

$$p_{\text{coin}} = \frac{1}{2}(1 + \text{Re}[\mathcal{V}e^{-i\delta}]), \tag{16.27}$$

from which $\mathcal{V}$ can be determined by taking measurements while sweeping through δ.

The problem with implementing the measurement is the difficulty of transporting the single photon state over long distances without incurring loss or additional phase shifts.

Instead of sending a valuable quantum state directly over a noisy quantum channel, one can distribute a Bell pair between the two telescopes; then we teleport the original quantum state from one telescope to the other. The entangled state is

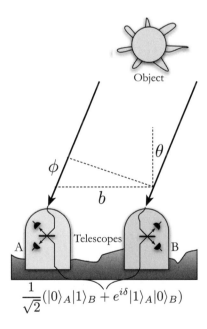

$$\frac{1}{\sqrt{2}}(|0\rangle_A|1\rangle_B + e^{i\delta}|1\rangle_A|0\rangle_B)$$

Figure 16.4 Architecture for quantum-enabled telescopy using two widely separated telescopes, which have shared Bell pairs. The basic idea is that the Bell pair mediates teleportation of one telescope's photon to the other, at which point an interferometric technique measures their phase difference, thereby determining ϕ. It is assumed that the baseline separation is large, $b \gg 1$.

known, the preparation can be repeated and one can use an entanglement distillation protocol to eliminate the phase noise.

Now, we can use the entangled pair directly to measure the visibility as in Figure 16.4. We post-select on the measurement results, considering the events where a single photon is observed at A and B simultaneously.

The variable delay line is now applied to the entangled state when the photon is sent to A, producing the entangled state

$$|\psi_{\text{shared}}\rangle = \frac{1}{\sqrt{2}}(|0\rangle_A|1\rangle_B + e^{i\delta}|1\rangle_A|0\rangle_B), \tag{16.28}$$

where δ is determined by a controllable delay, allowing completion of the protocol to determine ϕ.

Thinking futuristically, in a future large-scale quantum internet, whereby Bell pairs are a readily available resource across the globe, quantum-enabled telescopy need not be limited to pairs of telescopes but could expand to become large-scale telescope arrays comprising numerous telescopes, all sharing pairwise entanglement, distributed via the quantum internet. Using a 2D grid would enable θ to be measured along different axes, and the increased number of detectors would increase signal strength.

Part IV

Entanglement Distribution

17

Entanglement: The Ultimate Quantum Resource

As we have seen, the diversity of quantum states that may be communicated and protocols that are implemented over the quantum internet are extremely diverse, encompassing many different types of encodings and communications protocols.

Given this plethora of protocols and encodings, discussed in detail in Part III, one might ask whether there is a single primitive resource that might be applicable to all, or at least most, of these quantum protocols, thereby reducing the technological requirements of the nodes and quantum channels forming the network mediating them – if networks were able to specialise in a very limited number of tasks, we might reasonably expect them to be better optimised and exhibit better performance than a 'Jack of all trades, master of none' network!

It turns out that there is one particularly useful quantum resource that finds applicability in many of these protocols – *distributed entanglement*, which comes in many flavours and varieties, some of which we discuss now.

17.1 Bell States

Foremost, Bell pairs – the simplest entangled states – are an utterly indispensable resource for countless quantum protocols. In brief, Bell pairs find applicability in, amongst many others, the following key protocols:

- Cluster states: a Bell pair is also a two-qubit cluster state, a supply of which can be employed in fusion strategies to prepare larger cluster states, enabling universal, distributed measurement-based quantum computation.
- Quantum state teleportation: a shared Bell pair between Alice and Bob forms the elementary quantum resource upon which the state teleportation protocol is constructed.

- Quantum key distribution: the E91 QKD protocol is built upon a reliable stream of distributed Bell pairs, enabling private communication with perfect information theoretic security.
- Modularised quantum computation: using Bell pairs, entanglement swapping can be employed to fuse neighbouring but potentially distant modules together using operations local to each module.
- Superdense coding: a shared Bell pair enables the communication of two classical bits of information via transmission of a single qubit, thereby doubling classical channel capacity.
- Quantum-enabled telescopy: a shared Bell pair between two telescopes allows a photon received at one telescope to be teleported to the other, at which point interferometric techniques yield extremely sensitive phase information.

We see that Bell pairs form a ubiquitous resource, covering many of the most significant quantum protocols in quantum computation, distributed quantum computation, quantum state teleportation and quantum cryptography.

17.2 GHZ States

Beyond Bell pairs, multiqubit Greenberger–Horne–Zeilinger (GHZ) states (the direct generalisation of Bell pairs to n qubits) are useful in a variety of settings.

For the purposes of quantum anonymous broadcasting, multiparty GHZ entanglement is the primitive resource upon which the cryptographic protocol is constructed. As with Bell pairs, GHZ states are a known state and infinitely reproducible. They can also be purified. Thus, GHZ entanglement distribution is another useful primitive that future quantum hubs might specialise in preparing and distributing.

Additionally, quantum gate teleportation of a maximally entangling two-qubit gate (e.g., a controlled-NOT [CNOT] or controlled-phase [CZ] gate) is mediated via a shared four-qubit GHZ state. In a distributed environment the sharing of such a state between two parties (two qubits per party) enables implementation of a truly distributed two-qubit entangling gate.

17.3 Cluster States

Finally, cluster states are a primitive resource for measurement-based quantum computation. Owing to their handy ability to fuse together to one another, forming larger clusters, the preparation and distribution of relatively small cluster states lends itself well to distributed implementation by specialised providers. Providers could distribute small cluster states, which are subsequently fused together using simple two-qubit entangling operations to form desired topologies, either held locally or distributed in the cloud.

17.4 Why Specialise in Entanglement Distribution?

These observations warrant special treatment of entanglement distribution as a fundamental building block in the quantum era. One might envisage a quantum internet in which a central server(s), who specialises in only entangled state preparation and distribution, serves the sole role of pumping out Bell pairs or other entangled states across the quantum internet to whomever requests them, who subsequently use them for protocols such as those mentioned above. This could be in the form of a server transmitting over fibre networks or across free-space or via a satellite in orbit, transmitting at an intercontinental level.

What is the advantage of this approach to quantum networking? Why specialise in entanglement distribution, rather than implementing more capable networks with the ability to perform arbitrary operations? There are numerous reasons:

1. Dedicated servers can specialise in this one particular task, as can the transmission infrastructure.
2. The entanglement servers are entirely passive, not involved interactively with clients.
3. The server need not concern itself with the nitty-gritty of the protocols implemented by the end-user. It acts purely as a provider of a single resource, remaining uninvolved in their subsequent applications.
4. Because servers are providing a single standardised product, they can be commodified, enabling mass production of the hardware devices and the associated economy of scale. For example, mass production of simple ground-based Bell state relays or the construction of a comprehensive globe-enveloping constellation of satellites would inevitably improve economies of scale.
5. Unlike generic quantum states, Bell pairs, GHZ states and cluster states are known states that are infinitely reproducible, without having to worry about no-cloning limitations.
6. Photonic Bell pairs are easily prepared via type II spontaneous parametric down-conversion (SPDC) at very high repetition rates ($\sim$100 MHz–1 GHz), enabling rapid state preparation.
7. Small entangled states like Bell pairs are relatively 'cheap' to prepare and can be readily manufactured using widely accessible, present-day technology that has already been well-demonstrated on Earth and in space.
8. Quality of service is a lesser issue in most scenarios. We can employ a SEND-AND-FORGET protocol for the distribution of entanglement (much like classical

User Datagram Protocol): because every state is identical, we need not be concerned about missing ones. Instead, we can simply wait for the next one (a REPEAT-UNTIL-SUCCESS strategy), knowing that it will be exactly the same. We also call this the SHOTGUN approach – keep firing away until we hit something, and if we lose a few, who cares?

9. Rather than transmitting quantum states between distant parties directly, if we instead use state teleportation mediated by Bell states, the state to be transmitted will not be corrupted if the communications channel fails (e.g., via loss). Instead, we can wait for the next successfully transmitted Bell pair until we are ready to teleport the state, which then proceeds without directly utilising the quantum communications channel, accumulating its associated costs, or risking losing the state altogether should link failure occur. Only classical communication is required to complete the protocol, which can be regarded as error-free for all intents and purposes.

10. Entanglement purification may be employed by parties to improve the cost metrics associated with their shared entanglement, thereby partially overcoming the limitations imposed by the quantum communication channels.

11. If no direct link exists between server and clients, bootstrapped entanglement swapping can be employed to concatenate servers to create longer-distance 'virtual' links. This is the basis for *quantum repeater networks*.

17.5 Why Not Distributed Entangling Measurements?

In addition to entanglement distribution, entangling measurements (e.g., Bell state projections) may be used as a primitive for many protocols. This is effectively entangled state distribution in reverse, whereby two clients transmit states to a host, who performs a joint entangling measurement upon them. For example, in the modularised model for cluster state quantum computing, two adjacent but distant modules might transmit optical qubits to a satellite, which projects them into the Bell basis, thereby creating a link between the respective modules via entanglement swapping. This is not as powerful as entangled state distribution, because it cannot be used for, for example, E91 QKD, but nonetheless remains a powerful primitive for many protocols.

So which ought our quantum hubs specialise in, entanglement distribution, entangling measurements or both? For most practical purposes the former is far more powerful and robust. Let us take the example of fusing two remote cluster states together to form a larger, distributed virtual cluster. Imagine that their fusion operations are optically mediated by a satellite overhead. The options for satellite-mediated state fusion are as follows (see Figure 17.1):

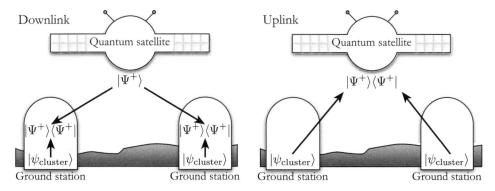

Figure 17.1 Satellite-mediated cluster state fusion operations for creating a link between two cluster states held by distant ground nodes. (left) Via entanglement distribution over a downlink channel. (right) Via distributed Bell projection over an uplink channel. When performing a distributed Bell measurement it is essential that the optical qubits arrive synchronously at the entangling measurement device, denoted $|\Psi^+\rangle\langle\Psi^+|$ (e.g a PBS), which is technologically challenging to implement on satellite given the unpredictable nature of the atmospheric quantum channel, necessitating on-board quantum memories to synchronise the qubits. This is likely to make downlinks cheaper, faster and more efficient than uplinks.

- Downlink mode: the satellite uses the downlink to distribute an entangled Bell pair between the two nodes. Each node performs a Bell projection between their half of the Bell pair and their respective qubit from their local cluster state, thereby swapping the entanglement and creating a link.

- Uplink mode: each node takes an optical qubit from their cluster state (or entangles their cluster state qubit with an optical qubit), which is uplinked to the satellite. The satellite performs a Bell projection between the two received optical qubits, thereby implementing entanglement swapping between the two nodes, creating a link.

Mathematically, these two processes are almost identical in their operation, differing only in direction. However, the former has the key advantage that it requires no time synchronisation operations on the server side, whereas the latter does, and satellite-based hardware is orders of magnitude more expensive than Earth-based hardware.

Both scenarios involve Bell projections. These entangling measurements require active synchronisation to ensure that the measured qubits arrive at the entangling measurement device (typically a polarising beam splitter [PBS]) simultaneously, to achieve high Hong-Ou-Mandel (HOM) visibility, requiring synchronisation on the order of the photons' coherence length. This can be achieved either using a brute-force REPEAT-UNTIL-SUCCESS mode of operation (post-selecting on events where both qubits arrive within a required temporal window) or storing one qubit in quantum memory until the other arrives. However, post-selection is expensive, requiring a massive overhead in the number of trials, and quantum memory is technologically challenging to implement, more so in space.

In the former case, the time ordering of the Bell projections performed locally on the ground nodes is irrelevant, although within each ground station the two qubits being projected must be synchronised, requiring quantum memories within ground stations.

On the other hand, in the latter case it is essential that both optical qubits arrive at the satellite's entangling measurement device simultaneously, which is extremely difficult to enforce when our quantum channels are tracking moving targets in low-Earth orbit and traversing a turbulent atmospheric channel in between. An on-satellite quantum memory would be extremely costly!

This yields several key advantages in favour of entanglement distribution as opposed to server-side joint entangling measurements:

1. The challenging prospect of quantum memory may operate on Earth, far less onerous and expensive than incorporating this technology into a satellite in low-Earth orbit.
2. Because the server is not storing any qubits in quantum memory, it does not suffer downtime associated with the periods between receiving the first photon and waiting for the second – it can continue to spit out Bell pairs at maximum capacity.
3. The satellite remains passive, implementing only the simplest of possible operations, reducing mass production costs.
4. The satellite does not require any interaction with its clients (classical or quantum).
5. Because Bell pairs are known, infinitely reproducible states, the server can operate in a User Datagram Protocol–like mode and it is not problematic if any given pair was lost. In the reverse direction, loss of a qubit could compromise the entire peripheral state associated with it in the ground station.
6. Entanglement purification can be employed to enhance the effective quality of the transmission channel.

We therefore anticipate that distributed entangling operations are likely to be mediated via entanglement distribution rather than distributed entangling measurements in the future quantum internet.

These observations lead us to naturally conclude that a quantum network specialised to this one particular task – entanglement distribution – would already be immensely useful and on its own enable many key applications.

18

Quantum Repeater Networks

In the previous section we concluded that quantum networks specialising purely in entanglement distribution (Bell pairs in the simplest case) would already be extremely capable in enabling many distributed quantum protocols. This motivates the development of protocols for entanglement distribution over noisy, long-distance quantum networks.

Any useful future quantum internet is going to require the communication of quantum information over arbitrarily long distances. Whereas intercity communication might be implemented via point-to-point connections, intracity and intercontinental communication will require extremely long-distance links, well beyond the attenuation length of the optical fibres connecting them or the line-of-sight of satellites in orbit.

Quantum repeaters [77, 158, 118] are devices that allow high-quality entanglement to be shared between distant nodes, when no direct line of communication is available from a server to its two clients. This is achieved by dividing long-distance links into a finite number of segments interspersed with repeaters (see Figure 18.1).

For example, a satellite in low-Earth orbit may be outside simultaneous line-of-sight to two distinct ground stations, owing simply to the curvature of the Earth. But this can be overcome by relaying a channel through several satellites in line-of-sight of one another. This is achieved using a bootstrapped entanglement swapping and purification protocol (Sections 16.5 and 16.2). Most commonly, this entanglement is in the form of Bell pairs, which, as discussed in Chapter 17, form a ubiquitous resource for many essential quantum protocols. The actual physical encoding of the entangled states may vary but is most commonly and archetypically in the form of polarisation-encoded single photons or continuous-variable states.

The links are now over much shorter distances and so can be generated with far higher probability. Then by stitching these together using entanglement swapping (Section 16.5), we can generate our required long-range entanglement link.

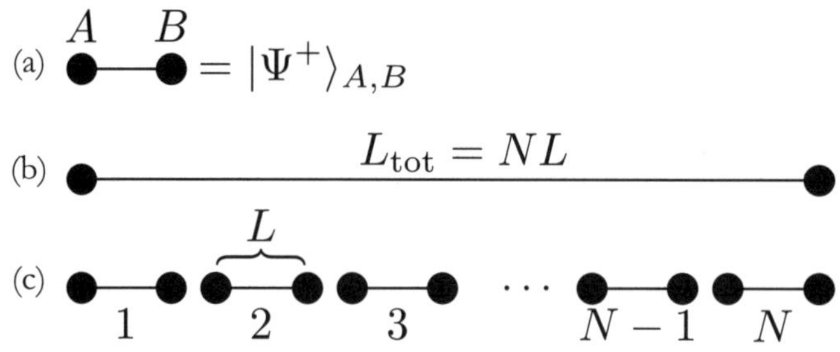

Figure 18.1 (a) Schematic representation of an entangled Bell pair $|\Psi\rangle_{A,B}$ shared between remote parties Alice and Bob. The two solid dots represent physical qubits, and the edge represents entanglement. (b) The link may be over a long distance L_{tot}. (c) Due to channel losses, the link may be broken into N smaller segments of length L. The links for each of the smaller segments can be independently generated and combined to form the longer distance link.

Beginning from this simple principle, the field of quantum repeater networks has grown enormously, leading to several generations of repeater designs, of ever increasing power and sophistication and ever more challenging technological demands.

18.1 First-Generation Repeaters

The above description is very hand-wavy, and of course things are a little more complicated in practise. We now examine these ideas in a little more detail, starting with a simple linear chain of repeater stations.

In a quantum repeater network, there are three main operations required:

1. Entanglement distribution: to create entangled links between adjacent repeater nodes.
2. Entanglement purification: to improve the quality of entanglement between nodes.
3. Entanglement swapping: to join adjacent entangled links together to form longer distance links.

The basic operation of a repeater, as shown in Figure 18.2, works as follows.

We begin our preparation of a long-range entangled link by creating multiple entangled pairs between adjacent repeater nodes (the number will depend both on the quality of the pairs we initially generate and on the target quality we want our final pair to have). Once we have enough pairs established between two repeater nodes, we perform entanglement purification, which converts multiple entangled links (pairs) into a smaller number with higher quality.

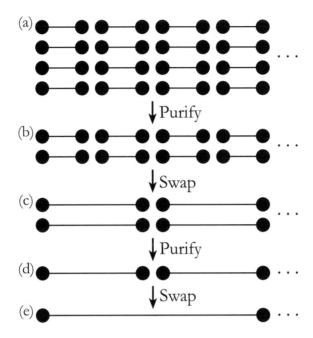

Figure 18.2 Basic operation of a (first-generation) quantum repeater network. (a) Preparation of multiple entangled links between adjacent repeater nodes. (b) They are then purified to create higher fidelity links. (c) Entanglement swapping between adjacent pairs creates links of twice the original length. (d) These new links are purified to create higher fidelity ones. (e) Entanglement swapping creates a link four times the original size. This process continues as necessary to reach target distance and purity.

These purification steps, shown in Figure 18.2(a) and 18.2(b), are performed on the links between all adjacent repeaters, increasing the quality of the links between those adjacent repeater stations to the required degree. Entanglement swapping, as shown in Figure 18.2(c), then creates links twice as long. The resulting entanglement links can then be used iteratively for further rounds of purification and swapping until one generates a high-quality link between the desired points in the network.

Entanglement Distribution

Probably the most important operation for any quantum repeater setup is entanglement distribution, the process of creating entanglement between two remote parties (Alice and Bob) connected by a quantum channel (generally an optical fibre or free-space link). This can be implemented in a number of ways [22, 64, 21, 158, 43, 181, 120] but can be broadly categorised into three basic schemas:

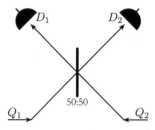

Figure 18.3 Entanglement distribution scheme based on quantum emitters and which-path erasure. Each node emits a photon entangled with the quantum memories present within that nodes. The photons from the adjacent repeater nodes then interfere on a beam splitter (or polarising beam splitter [PBS]), which erases information about which path the photon took. The photons are then measured in an appropriate basis to project the quantum memories within the nodes onto an entangled state.

- Photon emission from quantum memories in the repeater nodes, followed by which-path erasure.
- Absorption of entangled photons by quantum memories.
- Photon emission at one node and absorption at another.

By far, the emission based schemes are the most common, which we will concentrate on here. Such schemes operate by using an entangling operation – *which-path erasure* – to entangle two quantum memories via photons to which they were coupled. Effectively the process teleports the action of an entangling gate acting on the photons onto the quantum memories to which they were entangled.

We now describe such a which-path entangling operation in the context of two-level quantum memories coupled to polarisation-encoded photons.

Ideally one wants to initially generate a maximally entangled state of the form [118]

$$|\Psi\rangle = \frac{1}{\sqrt{2}}(|g\rangle|H\rangle + |e\rangle|V\rangle) \tag{18.1}$$

within the repeater node, where $|g\rangle$ and $|e\rangle$ are the two states (ground and excited) of the quantum memory and $|H\rangle$ and $|V\rangle$ are the polarisation states of a single photon. The photons from the two repeater nodes (Figure 18.3) are then transmitted to a beam splitter (or PBS in this example), after which the state of the system is

$$|\Psi\rangle = \frac{1}{2}|g\rangle|g\rangle|H\rangle|H\rangle + \frac{1}{2}|e\rangle|e\rangle|V\rangle|V\rangle$$

$$+ \frac{1}{2}|g\rangle|e\rangle|HV\rangle|0\rangle + \frac{1}{2}|e\rangle|g\rangle|0\rangle|HV\rangle. \tag{18.2}$$

One immediately notices that the $|g\rangle|e\rangle$ and $|e\rangle|g\rangle$ contributions are associated with two photons in one of the PBS exit modes, the other being in the vacuum state. However, the $|g\rangle|g\rangle$ and $|e\rangle|e\rangle$ terms have one photon in each of the output modes. They are of opposite polarisation, but measuring those photons in the diagonal/anti-diagonal ($\hat{X}$) basis erases this 'which-path' information, yielding an equal superposition of the two alternative histories – an entangled Bell state of the form

$$|\Psi_{\pm}\rangle = \frac{1}{\sqrt{2}}(|g\rangle|g\rangle \pm |e\rangle|e\rangle), \tag{18.3}$$

where the sign is given by the parity of the two photodetection outcomes in the $\hat{X}$ basis. This entangled state is stored in the quantum memories between nodes.

The scheme based on photon absorption by the quantum memories is effectively the time reversal of the emission-based scheme. Instead of using the beam splitter to entangle the photons emitted from each memory, a source of entangled photon(s) is employed. Of course, the emission and absorption schemes can be used together in a hybrid architecture.

In any entanglement distribution scheme for quantum networks, the repeater nodes are spatially separated and one must consider channel losses, which are the dominant error source. Channel loss in this situation implies that we do not register a coincidence event between D_1 and D_2, which heralds the entanglement. Thus, our entanglement distribution success probability is reduced. In fact, the heralded probability of success can be expressed as

$$p_{\text{ED}} = \frac{1}{2}e^{-L/L_0}p_{\text{det}}^2, \tag{18.4}$$

where L is the distance between the two repeater nodes, with L_0 being the attenuation length of the channel, and p_{det} is the detector efficiency. Here we have ignored the source and coupling efficiencies. It is immediately obvious from this expression that the further the repeater nodes are apart, the lower the probability of success, on an exponentially decaying trajectory. The attenuation length of typical telecom optical fibre is approximately 22.5 km and so the average time to generate a distributed entangled pair is

$$T_{\text{av}} \sim \frac{L}{c \cdot p_{\text{ED}}}$$

$$= \frac{2Le^{L/L_0}}{c \cdot p_{\text{det}}^2}, \tag{18.5}$$

where c is the speed of light in the channel. This grows exponentially against node separation and so places important constraints on the lifetime of the quantum memories. If we consider pure dephasing effects on our matter qubits, the state of our system can be represented by

$$\hat{\rho}(F) = F|\Psi^+\rangle\langle\Psi^+| + (1 - F)|\Psi^-\rangle\langle\Psi^-|, \tag{18.6}$$

where F is the fidelity of our entangled state given by

$$F = \frac{1 + e^{-t/\tau_D}}{2}, \tag{18.7}$$

where t is the duration over which the entangled state is held in memory and τ_D is the coherence time of the memory. If one only requires a single Bell pair and no further operation are performed, then $t = c/L$. However, in a more general setting where multiple pairs are required, the time will be T_{av} on average, which is inversely proportional to the probability of generating the entangled state. The quality of the prepared remote entangled state may therefore not be sufficient for the tasks it is required for due to these finite memory lifetimes or operational gate errors. One needs to be able to purify these entangled resources.

Entanglement Purification

The finite coherence time of quantum memories and operational errors caused by quantum gates means that some mechanism will be required to improve the fidelity of the distributed entangled state, especially if the spatial separation is large. This is generally achieved by entanglement purification [22, 51, 62, 131, 61, 7, 92, 121, 171], which as it name implies, purifies the entanglement to a higher value. The purification operation uses either an error detection code (probabilistic but heralded operations) [22, 51, 62] or deterministic error correction codes [7, 92, 121]. Though the error correction codes purify in a deterministic way, they place tough constraints on both the required initial fidelity of entangled states and the quality of the quantum gates implementing the purification [7]. Given this, we will focus on the simplest error detection code, which requires only a pair of shared entangled quantum memories (as shown in Figure 18.4). This scheme is equivalent to the entanglement purification protocol described in Section 16.2, although the graphical notation is somewhat different.

In this simplest purification protocol, Alice and Bob share two pairs of entangled states of the form given by Eq. (18.6). These states are a mixture of only two Bell states. We begin our purification protocol by using local operations to transform $\hat{\rho}$ to

$$\hat{\rho}(F) = F|\Psi^+\rangle\langle\Psi^+| + (1 - F)|\Phi^+\rangle\langle\Phi^+|. \tag{18.8}$$

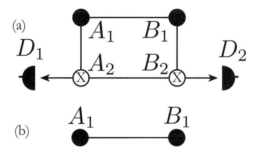

Figure 18.4 Entanglement purification: (a) The simplest purification scheme involving two pairs of shared remote entangled quantum memories ($A_1 - B_1$ and $A_2 - B_2$). The purification operation begins with Alice performing a controlled-NOT (CNOT) operation between memories A_1 and B_1. Similarly, Bob performs a CNOT operation between his memories. Alice and Bob then measure qubits A_2 and B_2 in the computational (0, 1) basis and share their results. They discard the resulting state if between them they measured odd parity (0, 1 or 1, 0). They keep the state if they measured an even parity between them (0, 0 or 1, 1), which should have higher fidelity. (b) Two qubits are removed, leaving a residual two-qubit state between A_1 and B_1 with improved fidelity.

As shown in Figure 18.4, we then apply a controlled-NOT (CNOT) gate between Alice's two memories and Bob's two memories following by measuring A_2, B_2 in the computational basis. Upon measurement of even parity, our resulting state $\hat{\rho}(F')$ has the form

$$F' = \frac{F^2}{F^2 + (1 - F)^2}. \tag{18.9}$$

It is immediately obvious that our resulting state $\hat{\rho}(F')$ is more entangled than $\hat{\rho}(F)$ when $F > 1/2$ (see Figure 18.5). In fact, the degree of entanglement as measured by the concurrence increases from

$$C = 2F - 1 \tag{18.10}$$

to

$$C' = 2F' - 1$$
$$= \frac{2F^2}{F^2 + (1 - F)^2} - 1. \tag{18.11}$$

It is important to mention that the entanglement purification does not allow one to distribute a *perfect* Bell state. Rather, it *asymptotically* approaches perfection (under ideal conditions) with repetition of the protocol.

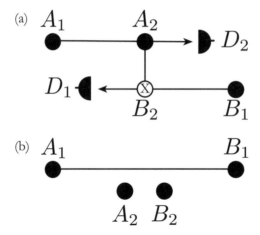

Figure 18.5 Plot of the increased fidelity and success probability for entanglement purification for a mixture of two Bell states with initial fidelity F. The dashed lines show how multiple pairs with an initial fidelity $F = 0.7$ can be purified iteratively to a final fidelity above 0.95.

The probability of obtaining the even parity outcome is

$$p_{\text{even}} = \frac{F^2 + (1 - F)^2}{2}. \tag{18.12}$$

Alternatively, for the odd parity measurement results, which occur with probability

$$p_{\text{odd}} = F(1 - F), \tag{18.13}$$

the resulting state is an equal mixture of $|\Psi^+\rangle$ and $|\Phi^+\rangle$ and is not entangled at all. In this case we must start again from scratch with the entanglement distribution.

So far we have discussed one round of entanglement purification, but the protocol naturally works in a recursive way where two copies of a state with the same fidelity are used for the next purification round. Using this bootstrapped approach one can in principle generate a near-unit-fidelity entangled pair from a finite-fidelity pair (provided initial input fidelity $F > 1/2$).

There are two common variants of these purification protocols: the Deutsch and Dür variants:

- *Deutsch protocol* [51]: This is an efficient purification protocol utilising Bell diagonal states that reaches a high fidelity in a few purification rounds. It is assumed that both entangled pairs have the same form. The purification protocol is the same at the one described in Figure 18.4 but begins with Alice (Bob) applying $\pi/2$ ($-\pi/2$) rotations about the X-axis on their qubits before the usual

CNOT gates and measurements are performed. Two copies of the successfully purified pair can then be used in a recursive approach to purify either further. This in turns means that multiple copies of the originally distributed states are required. We must have enough entangled pairs available to perform the multiple rounds of purification that are required, which grows exponentially with the number of purification rounds.

- *Dür protocol* [62]: This uses the same core purification elements as shown in Figure 18.4 but relaxes the traditional constraint that both Bell pairs must have the same fidelity. Instead, we begin with two pairs of the same fidelity F and perform the traditional purification. If successful, we perform the next round of purification using the improved fidelity pair from the previous round and a fresh fidelity F pair. In effect this new auxiliary pair is used to boost the fidelity of the original pair higher. This can continue until we reach a limiting fidelity dependent on the original F. This limiting fidelity may be above the desired resultant fidelity, at which point we can terminate the purification protocol. A significant difference between the Deutsch and Dür protocols is that the number of memories in the Dür situation is linear in the number of nesting levels.

It is critical in repeater protocols to also discuss how fast these purification protocols can be performed. Even with ideal gates one has to wait for the parity information to be shared between the repeater nodes. For nodes separated by a distance L, the communication time for a single trial is L/c. However, remembering that purification is probabilistic but heralded in nature, our waiting time could be many multiples of L/c. This will have a dramatic effect on performance, especially if performed at many different stages in the network with increasing distances between nodes.

Entanglement Swapping

The entanglement distribution and purification scheme discussed previously allow one in principle to create high-fidelity entangled states between adjacent repeater nodes. The next task is to extend the range of our entangled states, and this occurs via simple entanglement swapping [35, 194, 78, 59]. This was described previously in Section 16.5, although the notation is modified.

Consider the situation where we have an entangled Bell pairs between nodes A_1 and A_2 and also between B_2 and B_1. The entanglement swapping operation involves a Bell state measurement between the qubits A_2 and B_2 as shown in Figure 18.6. After the Bell measurement, we have the resultant state

$$\hat{\rho}_{A_1, A_2}(F) \otimes \hat{\rho}_{B_2, B_1}(F) \rightarrow \hat{\rho}_{A_1, B_1}(F') \tag{18.14}$$

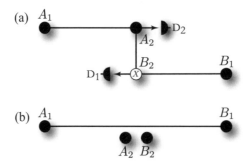

Figure 18.6 Entanglement swapping: (a) An entangled state is shared between Alice and a repeater station ($A_1 - A_2$) and also between the repeater station and Bob ($B_2 - B_1$). The entanglement operation begins by performing a Bell state measurement between A_2 and B_2 using a CNOT gate and measurements at D_1, D_2. The measurements indicate which Bell state we have projected our state A_1, B_1 onto. (b) The resultant entangled state between A_1, B_1 with the qubits A_2 and B_2 disentangled from it.

with

$$F' = F^2 + (1 - F)^2, \tag{18.15}$$

where a local correction operation is performed on either A_1 or B_1 depending on the measurement outcome. It is clear that the longer range entangled state $\hat{\rho}_{F'}$ is less entangled that the states $\hat{\rho}_F$ used to generate it. In fact, to first order our fidelity drops from F to F^2. This in turn means that we cannot simply purify adjacent repeater pairs and swap them all to create the long-range pairs. If we had n links, our final fidelity from all of the swapping would scale as F^n. For high-fidelity end-to-end entangled links we need to follow the approach outlined in Figure 18.2. Finally, depending on how the Bell measurement is implemented, this process could be probabilistic (but heralded) or deterministic in nature. We assign the success probability as p_{ES}.

Performance

We now have all of the operations required for a repeater to create long-range entanglement. The natural question to ask is how well it performs.

There are several important points to initially consider here. The majority of the repeater operations are probabilistic in nature (entanglement distribution and purification fundamentally and entanglement swapping dependent upon implementation). Though these probabilistic operations may be heralded, classical signalling must be performed between involved nodes to inform them of successes or failures. Entanglement distribution this time is just that associated with the signalling between adjacent nodes. However, purification and swapping are likely to require such signalling over the entire length of the network. This has a dramatic effect on the performance of the repeater network. The normalised rate for generating Bell pairs over a total distance L_{tot} is given by

$$R(n, k, L_{\text{tot}}) = \frac{1}{T_{n,k,L_{\text{tot}}} M_{n,k}}, \tag{18.16}$$

where $T_{n,k,L_{\text{tot}}}$ is the time to generate a Bell pair over the total distance using an n-nested repeater configuration with k rounds of purification per nesting level. The distance between repeater nodes is given by

$$L = \frac{L_{\text{tot}}}{2^n}, \tag{18.17}$$

meaning that there are $2^n - 1$ intermediate repeater nodes with Alice and Bob at the endpoints. In Eq. (18.16) we discount our rate by $M_{n,k}$, the total number of quantum memories used. The justification for this is that this provides a fairer comparison when different purification approaches are used. The Deutsch protocol, for instance, achieves its target fidelity much faster (fewer rounds) than the Dür protocol but consumes far more resources in doing so.

Now it is straightforward, albeit tedious, to show that $T_{n,k,L_{\text{tot}}}$ is given by [32],

$$
\begin{aligned}
T_{n,k,L_{\text{tot}}} &\sim \frac{3^n}{2^{n-1}p_{\text{ED}}} \prod_{i=0}^{n-1} \left(\frac{3}{2}\right)^k \frac{1}{P_{\text{ES}}(n-i)} \prod_{j=0}^{k-1} \frac{1}{p_{\text{P}}(k-j,n-i)} \\
&+ \sum_{m=1}^{n} \left(\frac{3^{n-m}}{2^{n-1}}\right) \prod_{i=0}^{n-m} \left(\frac{3}{2}\right)^k \frac{1}{P_{\text{ES}}(n-i)} \prod_{j=0}^{k-1} \frac{1}{p_{\text{P}}(k-j,n-i)} \\
&+ \sum_{m=1}^{n} \sum_{q=0}^{k-1} \left(\frac{3^{n-m+q}}{2^{n-2m+q}}\right) \prod_{r=0}^{q} \frac{1}{p_{\text{P}}(k-r,m)} \prod_{i=0}^{n-m-1} \left(\frac{3}{2}\right)^k \frac{1}{P_{\text{ES}}(n-i)} \\
&\times \prod_{j=0}^{k-1} \frac{1}{p_{\text{P}}(k-j,n-i)}, \tag{18.18}
\end{aligned}
$$

where p_{ED} is the probability of successfully distributing entanglement between adjacent repeater nodes, and $p_{\text{P}}(j,i)$ [$p_{\text{ES}}(i)$] represents the purification (entanglement swapping) probability at the ith nesting level with j rounds of purification. The factor of $3/2$ present in all entanglement distribution, purification and swapping operations is a multiplicative factor associated with the extra time required for the two pairs to be available for the various quantum operations [158].

It can be easily seen from this formula that

$$T_{n,k,L_{\text{tot}}} \gg \frac{2L_{\text{tot}}}{c}, \tag{18.19}$$

especially if probabilistic gates are included. Next the resources scale polynomially with

$$M_{n,k} \sim 2^{(k+1)n}$$

$$= \left(\frac{L_{\text{tot}}}{L}\right)^{k+1} \tag{18.20}$$

for the Deutsch protocol, which in turn implies that it is efficient. However, for long distances L_{tot}, our normalised rate $R(n,k,L_{\text{tot}}) \ll 1\text{Hz}$, especially when probabilistic CNOT gates and Bell state measurements are employed [92, 119].

18.2 Second-Generation Repeaters and Error Correction

The previous approach for entanglement distribution over long distances based on first-generation quantum repeaters has its performance heavily constrained by both the probabilistic nature of the various quantum operations and the associated classical communication time. We know that the classical communication in entanglement distribution is only between the adjacent nodes, whereas for the purification and swapping operations it can be very long-range, potentially over the entire network length. This is the fundamental reason why the time to create a pair is of order $O(L_{\text{tot}}/c)$ or longer. This will not change significantly even if we have deterministic CNOT gates and Bell measurements because the entanglement purification protocols will remain probabilistic in nature (even though the swapping operations will be deterministic). We thus need to replace our usual entanglement purification protocols with a similar operation that is deterministic in nature [92, 119].

The typical entanglement purification protocols are a form of quantum error detection code [118, 53]. Such codes herald whether an error has occurred or not, and in the situation considered above, detection of errors means that one must discard the entangled pairs associated with the purification protocol. No errors means that the purification protocol has worked.

Error correction codes that operate in a deterministic fashion can also detect errors and can be used in this fashion [92, 119]. More critical, quantum error correction codes have the potential to correct some errors that have occurred, mitigating the need to completely discard states affected by errors. For normal error correction protocols used in quantum computations, we encode our physical qubits into logical

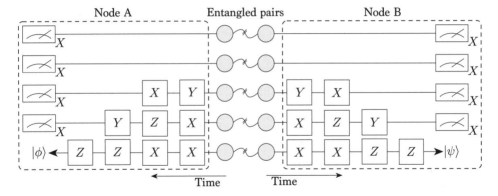

Figure 18.7 Purification circuit based on quantum error correction. The specific example shown is for the [5,1,3] code [24, 103]. We assume that entanglement distribution has allowed Alice and Bob to create five copies of their imperfect Bell pairs. The error correction circuit is executed independently between the two nodes. Though we show the situation when the measurements at both sides are done directly on four pairs of entangled qubits (leaving us with one unencoded Bell pair), one can also use ancilla qubits to measure the appropriate syndromes. As soon as the measurements are complete, both nodes' qubits are available for continued use because the error correction is deterministic and there are no failure events that need to be heralded. In this case the classical message between nodes just carries Alice's measurement results, allowing either node to interpret which Bell state was generated and for one of them to apply the bit-flip or phase-flip correction operation if needed to recover the desired Bell state. In many cases this correction is classically tracked in the Pauli frame, which keeps a record of whether $\hat{X}$ and/or $\hat{Z}$ corrections need to be performed at some stage [92, 119]. Note that it is not necessary to measure out all but one of the qubits involved in the entangled links. Instead, the logical qubit can be maintained by the use of ancilla qubits within that node with the syndrome being measured with the help of the ancilla qubits. Entanglement swapping could then be performed on the logical qubits, enabling a much more error-resilient system.

qubits using the code and then use syndrome measurements to determine where an error has potentially occurred.

Quantum communication, however, is different in this case because we must assume that we have generated a number of imperfect Bell pairs between the repeater nodes before we utilise the error correction schemes. The error correction protocol in this case operates by using the error correction encoding circuit on Alice's qubits and the decoding circuit on Bob's [7] as illustrated in Figure 18.7 for the five-qubit code [24, 103].

It is important to state here that error correction–based purification is deterministic in nature (there is, however, a significant cost that must be paid – the fidelity of the originally generated entanglement between adjacent nodes must be quite high) [92, 7]. There are no measurement events that need to be discarded. Instead, the measurement results only inform us of which particular imperfect Bell states we have and the correction operation required to return to the desired state. In effect the measurement is updating the Pauli reference frame [102]. This does not need to be executed immediately and may be deferred until later. In turn, this means that once the measurements have been performed, we can immediately use the purified Bell state without having to wait for the classical signalling (at some stage the correction operation needs to be executed, but this can be once the long-distance entanglement has been generated).

Mitigating having to wait for the measurement results to be sent and received in both the quantum error correction–based purification and entanglement swapping protocols has a profound effect on the rate of generating long-range entangled pairs. We still need to perform long-range classical messaging (potentially between end nodes), and thus it is immediately obvious that the preparation time can scale solely as

$$T = \frac{2L_{\text{tot}}}{c}, \tag{18.21}$$

which was the lower bound on the first-generation schemes [119].

Naïvely, this seems to imply that the generation rate between end nodes cannot be faster than this. However, one can in fact do far better! This is shown in Figure 18.8 (protocol described in caption). The key issue is that the generation rate depends on how long the adjacent nodes need to store part of an entangled state [92, 119, 123].

Fundamentally we know that the time to attempt to generate a single entangled Bell pair between two nodes is scaling as L/c (where L is the distance between those two nodes). With channel losses we need to make

$$m = \frac{\log_{10}(\varepsilon)}{\log_{10}(1 - p_{\text{ED}})} - \log_{10}\left(\frac{\varepsilon}{p_{\text{ED}}}\right) \tag{18.22}$$

attempts to generate a single Bell pair with error probability ε. We can make these attempts simultaneously and not affect the generation time. Now by using a butterfly repeater design, as illustrated in Figure 18.8, one immediately notices that the qubits with the repeater nodes are only used for duration $\sim 2L/c$. After this time those qubits have been freed up and are available to generate new entangled links. This means in turn that the time to generate the long-range entangled pair will scale as $T \sim O(2L/c)$ (independent of the overall distance L_{tot}) [92, 119, 123]. The exact resources used depend heavily on the error correcting code, but we know that they in principle scale as $M \sim O(\text{polylog}(L_{\text{tot}}))$ [123].

This is quite a dramatic decrease in both T and M compared to the first generation. In fact, one could expect the normalised rates to be on the order of kilohertz [119]. However, this is a significant cost in terms of the quality of the original

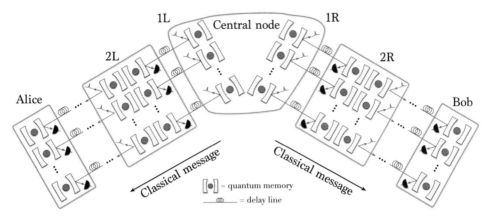

Figure 18.8 A butterfly design quantum repeater network protocol that reduces the requirements on all of the quantum memory times to only that associated with the signalling time between adjacent repeater nodes [119]. Enough pairs must be generated between the node to ensure that we can use them in the error correction code in a single round-trip time between adjacent nodes. The scheme relies on multiple entangled pairs being generated temporally, starting from the mid-point of the network. The protocol begins with the central node creating links to both the left and right nearest neighbour nodes in sufficient number to allow an error correction code to be implemented. Once they are created, the error correction circuits are applied to the links left and right of this central node (effectively creating encoded logical links). Entanglement swapping at the middle node is then applied between these logical links, creating a logical link between the left and right adjacent nodes. The left and right nodes can then do the same to their next adjacent repeater nodes, error correcting as they go, until the desired end-to-end entangled link is achieved.

Bell pairs that must be prepared. In the first-generation schemes a fidelity just over 50% was sufficient. However, with the second-generation schemes using normal error correcting codes, it is likely that this initial fidelity will have to be over 90% [92, 119].

18.3 Third-Generation Repeaters

The use of error correcting codes significantly improves the performance of second-generation quantum repeaters compared to first-generation ones. The second-generation schemes are now limited by the communication time between adjacent repeater nodes to herald whether entanglement distribution was successful or not [119, 121]. The communication (both quantum and classical) is ultimately limited by the speed of light (either in fibre or over free-space). The natural question is whether we can improve performance even further.

The only remaining avenue at our disposal is to move from probabilistic to deterministic entanglement distribution. Remembering that we have losses in the channel, the only way to achieve deterministic entanglement distribution will be by transmitting encoded error-correctable states between repeaters. This means that we must turn to loss-based error correction codes [136, 121, 68, 9, 122].

Loss-Tolerant Codes

There are quite a number of error codes that can correct for loss events, but here for illustration we consider *parity codes* in their simplest form [136, 121]. Other well-known approaches are based on cluster states.

Consider a four-photon state of the form

$$|\Psi\rangle = (\alpha|0\rangle_1|0\rangle_2 + |1\rangle_1|1\rangle_2) \otimes (|0\rangle_3|0\rangle_4 + |1\rangle_3|1\rangle_4)$$
$$+ \beta(|0\rangle_1|1\rangle_2 + |1\rangle_1|0\rangle_2) \otimes (|0\rangle_3|1\rangle_4 + |1\rangle_3|0\rangle_4), \qquad (18.23)$$

where $|0\rangle$ and $|1\rangle$ represent orthogonal degrees of freedom (e.g., polarisation). This state can be rewritten in the form

$$|\Psi\rangle = \alpha|\Phi_{12}^+\rangle|\Phi_{34}^+\rangle + \beta|\Psi_{12}^+\rangle|\Psi_{34}^+\rangle, \qquad (18.24)$$

and thus the state has been encoded into terms of a tensor product of two redundantly encoded Bell states. Now photon loss will remove one of these photons. As an example, let us consider what happens when photon 4 is lost. The resultant state can be represented by the density matrix

$$\hat{\rho} = |\zeta^+\rangle\langle\zeta^+| + |\zeta^-\rangle\langle\zeta^-|, \qquad (18.25)$$

where

$$|\zeta^+\rangle = \alpha|\Phi_{12}^+\rangle|0\rangle_3 + \beta|\Psi_{12}^+\rangle|1\rangle_3,$$
$$|\zeta^-\rangle = \alpha|\Phi_{12}^+\rangle|1\rangle_3 + \beta|\Psi_{12}^+\rangle|0\rangle_3. \qquad (18.26)$$

We immediately notice that $|\zeta^-\rangle = \hat{X}_3|\zeta^+\rangle$ and so by measuring the third photon in the $\hat{X}$ basis our state reduces to the pure state $\alpha|\Phi_{12}^+\rangle \pm \beta|\Psi_{12}^+\rangle$, where the $\pm$ sign is given by the $\hat{X}$ measurement outcome. This is then correctable using local operations. After the loss event of photon 4 and the measurement of the third photon, our state thus becomes

$$\alpha|\Phi_{12}^+\rangle + \beta|\Psi_{12}^+\rangle, \qquad (18.27)$$

which has exactly the same information in it as $|\Psi\rangle$ but without the redundant encoding.

It is now straightforward to re-encode back to our original state, Eq. (18.24). We considered photon loss only on the fourth qubit. However, the same principle applies for any lost photon. Unfortunately, we can only tolerate the loss of a single photon using this encoding, so the loss rate must be small.

The above example illustrates how the smallest optical loss code works. The general code with $n - 1$ redundancy can be written as [136, 121]

$$|\Psi\rangle = \alpha|\Phi_e\rangle_1 \ldots |\Phi_e\rangle_n + \beta|\Psi_o\rangle_1 \ldots |\Psi_o\rangle_n, \tag{18.28}$$

where $|\Phi_{e,o}\rangle$ are the even and odd parity m photon states given by

$$|\Phi_{e,o}\rangle = \frac{1}{\sqrt{2}}(|+\rangle_1 \ldots |+\rangle_m \pm |-\rangle_1 \ldots |-\rangle_m), \tag{18.29}$$

with $|\pm\rangle = \frac{1}{\sqrt{2}}(|0\rangle \pm |1\rangle)$. This redundancy-based parity code is composed of n logical qubits each containing m photons. For this code to correct loss errors we have two constraints:

- At least one logical qubit must arrive without photon loss.
- Every logical qubit must have at least one photon arrive successfully.

If these constraints are met, the loss events during transmission between adjacent repeater nodes can be corrected. Of course, such codes cannot correct more than 50% of errors and so the distance between repeater nodes is limited. Remembering that the probability of a photon being successfully transmitted through a channel of length L with attenuation length L_0 is given by $p = e^{-L/L_0}$, the maximum distance between repeater nodes is $L/L_0 \sim 0.69$ (which corresponds to approximately 17 km in present-day commercial telecom fibre). This is much shorter than what we would typically consider for the first- and second-generation schemes.

Operation

Let us now describe the operation of the third-generation repeater scheme depicted in Figure 18.9 in detail [121, 122].

It begins at the left-hand node by Alice encoding her message into a redundant parity code created on a series of matter qubits using local quantum gates within that repeater node.

The quantum state is then transferred/teleported via photons that are transmitted through a lossy channel to the adjacent repeater node. Here, two specific operations occur: first, the information encoded on each photon is transferred to a matter qubit within that repeater node and then that photon is measured. The photon measurement is critical because it heralds which photons have been lost and allows us to measure the remaining qubit in that block in the $\hat{X}$ basis, which removes the damaged parity blocks from our encoded state, leaving our information intact. We can now add the full redundancy back into our encoded state in the matter qubits. The fully encoded state can then be transferred to photons and transmitted to the next repeater node where the same procedure occurs again. This continues until our state reaches the last repeater node where Bob is.

There is one immediate observation that can be made from this scheme. The matter qubits (quantum memories) within the local nodes are only used to encode and error correct the redundancy code as well as transmit those quantum states as photons. Entanglement is not stored within the nodes while the photons are being sent to the adjacent repeater nodes. This in turn means that the resources within that repeater node can be used immediately again (once the photons have been transmitted), and so the rate of communication is now limited by the time to perform the local operations within a node, rather than the round-trip time between adjacent nodes.

The focus so far has been only on loss-based errors, but this code is fault-tolerant to general errors as well [122]. Furthermore, this redundancy code was only an illustrative example that photon loss in the channel can be corrected. Many other codes can be used in a similar fashion [121, 68, 122]. Finally, the scheme we present in Figure 18.9 transmits a quantum signal from Alice and Bob. It can, however, be

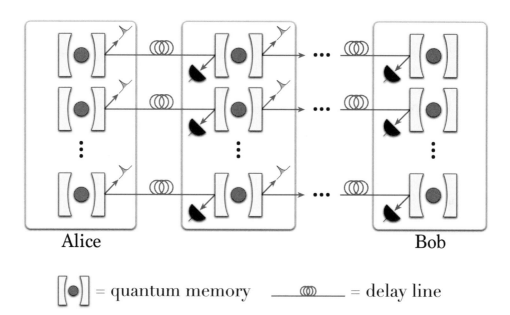

Figure 18.9 Transmission of a quantum signal using loss-based error correction codes in a quantum network.

adapted to use the butterfly design from Figure 18.8 to create remote entanglement between Alice and Bob while maintaining the performance advantages that our direct transmission scheme gave.

18.4 The Transition to Quantum Networks

The previous quantum repeater networks we discussed have been simple point-to-point linear networks. Though there may have been a number of ways to establish end-to-end entangled links between Alice and Bob, they knew that they were connected via a simple, direct, linear chain.

Of course, this is highly unrealistic. Alice and Bob are likely to be members of a complex quantum network that supports multiple users simultaneously and offers multiple routes from a given source to a given destination. This leads to a number of interesting considerations going forward:

- For large-scale networks the users may not know its exact network topology or even the best route between them. In fact, there could be multiple paths between Alice and Bob. Probing the entire network to establish the best route would be slow and costly (in practice). Still, every node should have a unique identifier (quantum IP address) that uniquely indicates its location and resource availability in the network.
- Most complex networks dynamically change in time as resources become congested or nodes break. This in turn means that using a butterfly approach to create Alice and Bob's links is problematic because one does not know the middle point between them to start the entanglement creation process. If one has to determine the route in advance and restrict access to those parts of the network required to establish the entire links, congestion will quickly follow. The generation rate will be very slow.
- Finally, it is unlikely that repeater nodes will be equally spatially separated (making the first-generation repeater schemes extremely hard to use in this situation).

The above issues lead us to a network model where Alice and Bob suspect that there is a route between them but do not know the exact route, which is likely to be dynamic; i.e., the availability of routes and their relative costs are liable to change. In such a case, if Alice wants to send a message to Bob, she uses her knowledge of Bob's rough location (from the quantum IP address) and her knowledge of the nodes close to her to send a message to a repeater node, who will have more knowledge of Bob's part of the network. This node can then forward the message to further nodes (who know even more about Bob's location) until it finally reaches Bob. The quantum IP address is essential here because that identifier indicates to the repeater node who to forward to next. In principle, as the message (or entanglement) is being established node by node, those repeater nodes who have already been used are free to work on tasks for other users.

There is another interesting aspect of our general complex quantum networks. There are likely to be many paths between Alice and Bob that could be attempted in superposition fashion. This will increase not only the capacity between Alice and Bob but also its robustness.

19

The Irrelevance of Latency

Entanglement distribution can be executed in a highly varying manner of ways – from transmitting optical qubits through space via satellites, to across land surfaces via optical fibre, to dumping solid-state qubits into cargo containers and shipping them via land or sea freight. These bring with them associated transmission latencies. The former two distribute entanglement at the speed of light with latencies on the order of microseconds, whereas the latter induces enormous latencies on the order of days or weeks.

At first glance it may appear that this renders the Sneakernet approach to entanglement distribution useless. Who wants to wait several weeks to communicate their qubits?

If these transmission methods were being utilised for direct transmission of quantum data, this would certainly be a major concern. However, we are not employing them to communicate unknown quantum data packets directly. Rather, we are using them to distribute many instances of completely identical Bell states. This changes the impact of latency entirely. That is, we treat known entangled states as a *resource* rather than as an actual unit of data, and provided that we can store it (i.e., we have a good quantum memory), whether it arrives sooner or later is not terribly important. More important is that we have a 'buffer' of entangled states at hand to draw upon when needed.

If our goal is to transmit a quantum state between two parties, the obvious approach is to send the qubits directly over the quantum channel. Alternatively, they could initially share Bell pairs and then employ quantum state teleportation to teleport the state between parties. In this case all that matters is that they hold a shared Bell pair in time for execution of the teleportation protocol. It could have been distributed between them at any point in the past and held in a quantum memory until needed. The latency is now determined entirely by the latency of the *classical* channel, which communicates the associated local corrections required to complete the teleportation protocol. In most classical networks, communication rates are on the order of the speed of light, with very little latency.

We see that the latency associated with entanglement distribution does not affect the latency of quantum state transmission when implemented via teleportation. The quantum network could continually be sharing entangled pairs between parties in a User Datagram Protocol–like mode, which holds them in quantum memory. They ensure that Bell pairs are being distributed at a sufficient rate that parties have a buffer of entangled pairs sufficient to accommodate demand for future teleportations. This irrelevance of quantum latency is a uniquely quantum phenomena, not applicable to any classical protocols.[1]

Teleportation-based quantum communication is additionally favourable in that shared Bell pairs can be purified before being utilised, allowing errors accrued during quantum communication to be minimised, something not so straightforward (or impossible) when transmitting data qubits directly.

---

[1] One minor exception might be to treat randomness as a resource for randomised classical computation; i.e., for application in **BPP** algorithms. In that restricted instance the latency of our source of random bit strings is also irrelevant because randomness is invariant under temporal displacement and can be buffered for future use.

20

The Quantum Sneakernet

In the quantum Sneakernet entanglement distribution protocol, rather than using conventional flying qubits, we take highly error-corrected stationary qubits and physically transport them over long distances as freight. The error correction must be sufficient to maintain coherence over the timescale of the journey, which could be anything from hours (when flying) to weeks (by sea[1]). The mode of transport only transports one half of an error-corrected Bell pair, transferring the initial entanglement between source and vessel, $|\Psi^+\rangle_{A,B}$, to be between source and destination, $|\Psi^+\rangle_{A,C}$. Because all Bell pairs are identical and infinitely reproducible, latency presents no problem. For example, when using a Bell pair to teleport a qubit over long distances, when the Bell pair arrives is unimportant, as long as it is available at the time of teleportation. Thus, our Bell pair cargo carriers can simply operate in the background, transporting Bell pairs with as much bandwidth as possible, which may then be buffered by the recipient until required, because latency is unimportant provided that the coherence lifetime of the error correcting code is long enough.

---

[1] Or indefinitely if the destination is Australia and you are fleeing war crimes.

Part V

Quantum Cryptography

Undoubtedly, quantum technologies will be most impactful (and disruptive!) in the area of information security, something of fundamental importance to us all on a daily basis and vital to the entire world economy. Quantum technologies will be important in terms of both breaking and maintaining security, with the former mandating interest in the latter.

In Chapter 30 we discuss encrypted outsourced quantum computation as an important concept in future cloud quantum computing. In this section we will step back from full-fledged distributed quantum computation, instead focussing on more elementary protocols for simple secure communication or protocols.

Today, the ability to communicate secretly with others is completely taken for granted in all but a few nations and resides in every smartphone and desktop PC. Furthermore, the encryption technologies available to the average consumer are extremely strong – the same as those used by large organisations, including world governments.

21

What is Security?

Before describing any specific cryptographic protocols, let us define what is meant by 'security' in a cryptographic context. We differentiate between *information theoretic security* and *computational security*:

- Information-theoretic security: the laws of quantum information bound the amount of information that can be extracted from a system, irrespective of measurement or computational operations. Thus, such security can be regarded as attack-independent, making no assumptions about our adversary's capabilities.
- Computational security: is based on the assumption that an adversary's computational resources are insufficient to perform cryptanalysis or brute-force cracking.

Clearly the former makes a far stronger statement about the security of a protocol than the latter.

Classical public- and private-key encryption protocols are typically based on the assumption of computational security (e.g., the computational complexity of performing integer factorisation in the case of Rivest-Shamir-Adleman (RSA) public-key encryption or solving a complex satisfiability problem in the case of private-key encryption), whereas quantum encryption protocols are typically information theoretically secure (e.g., the one-time pad using quantum key distribution).

22

Classical Cryptography

We begin with an introduction into *classical* cryptography to understand its limitations, which logically leads us into how quantum mechanics can assist in overcoming them. We only scratch the surface of this extremely well-researched field, reviewing some of the most important and widely used protocols. For a deeper understanding of classical cryptography we refer the interested reader to the excellent and comprehensive [162].

22.1 Private-Key Cryptography

Private-key (or symmetric-key) cryptography is perhaps the most basic (and useful) cryptographic primitive, enabling encryption of a channel between two parties who share a secret key – a random bit string of length determined by the encryption algorithm. The same secret key is employed for both encryption and decryption operations (hence 'symmetric'), making it of utmost importance that it be retained secret.

Private-key cryptography has a long history, in fact going back to ancient times, enabling the secret sharing of diplomatic messages between emperors and empires; e.g., the so-called *Caesar cipher*, a simple substitution cipher based on shifting the letters of the alphabet. However, it was a niche technology that very few utilised, because it had to be implemented by hand without computers or automation.

Today there are countless freely available private-key cryptographic protocols available online, and some have been standardised by standards institutes. Currently, the Advanced Encryption Standard (AES) is a standard endorsed by the US government, replacing the earlier standardised Data Encryption Standard (DES) whose mere 56-bit key length is today considered insecure in light of present-day computing power. AES is a block cipher, meaning that it divides data into small blocks of 128 bits, each of which are encrypted independently, and operates with key lengths of up to 256 bits (referred to as AES256), making it very robust against (even quantum) brute-force attacks. The length of the plaintext and ciphertext is the same, meaning that there is no bandwidth overhead when communicating encrypted data across a network.

22.2 One-Time Pad Cipher

There is one and only one *provably* secure (in the sense of information-theoretic security as opposed to computational security) encryption protocol – the *one-time pad*. This protocol requires Alice and Bob to share a random secret key as long as the message (plaintext) being communicated between them. The two bit strings undergo bit-wise exclusive-OR (XOR) operations to form the ciphertext. Mathematically,

$$c = s \oplus k, \tag{22.1}$$

where $\oplus$ is the bitwise XOR operation (equivalently addition modulo 2) and c, s and k are the ciphertext, plaintext and key bit strings, respectively, all of which are of the same length,

$$|c| = |s| = |k|. \tag{22.2}$$

The security of this protocol is easy to see intuitively – with an appropriate choice of key, *any* plaintext of the same length could be inferred from *any* ciphertext. This means that there is no possibility of performing any kind of frequency analysis, because the ciphertext string has maximum entropy (inherited from the maximum entropy of the random key and assuming a strong cryptographic random bit generator) and thus no correlations. Because every possible valid plaintext can be recovered using an appropriate key, a cracking algorithm is unable to find a unique plaintext matching the ciphertext, because all are equally valid decryptions.

Importantly, the secrecy of the one-time pad strictly requires that a key never be reused. A fresh key must be generated for each message sent; otherwise, trivial frequency analysis techniques can be employed to compromise security. If the same key k is used to encode two messages s_1 and s_2, yielding ciphertexts

$$c_1 = s_1 \oplus k,$$
$$c_2 = s_2 \oplus k, \tag{22.3}$$

then we trivially obtain

$$
\begin{aligned}
c_1 \oplus c_2 &= (s_1 \oplus k) \oplus (s_2 \oplus k) \\
&= (s_1 \oplus s_2) \oplus (k \oplus k) \\
&= s_1 \oplus s_2, \tag{22.4}
\end{aligned}
$$

which is independent of the key. Now a frequency analysis on the bitwise XOR of two plaintexts can be applied, without requiring any knowledge of the key whatsoever.

Needless to say, the requirement for keys of the same length as the plaintexts, which cannot be reused, raises the obvious criticism that now secret key sharing is as difficult as sharing a secret message in the first place. This reduces the problem of perfect secrecy of arbitrary messages to the secrecy of shared randomness.

Although during the Cold War Soviet diplomats would literally carry briefcases between countries full of paper with random data for use in a one-time pad, it is clearly not suitable for everyday applications!

22.3 Public-Key Cryptography

Though private-key cryptography solves the problem of end-to-end cryptography, it has one main downfall: how does one share a private-key between two parties? After all, if we had the ability to secretly share keys between ourselves, would we not just use that same method to directly communicate, bypassing the unnecessary cryptographic protocol?

Public-key (or asymmetric-key) cryptography addresses this issue by replacing the private key with two keys (known as a key pair), one used solely for *encryption* and the other solely for *decryption*. Importantly, these two keys are nontrivially related and cannot be efficiently computed from one another. To send a message to a friend I can send him my encryption (public) key, which he is only able to use for preparing an encrypted message for me. No security is required when sharing the public key because an eavesdropper cannot use it for decryption. Finally, I am able to decrypt the message using my decryption (private) key, which I kept completely to myself and never shared with anyone.

Since RSA, numerous other public-key cryptosystems have been developed, based on different choices of trapdoor function. Most notable, elliptic-curve cryptography has gained much attention. However, RSA remains the most widely used and well-studied public-key cipher.

To mitigate the need for constant one-on-one exchange of public keys, many key servers exist around the globe, which maintain databases of people's public keys. These servers are in a position of trust, vouching for the identities associated with their stored public keys.

22.4 Digital Signatures

Rather than cryptographically ensuring the secrecy of messages, a user may wish to prove their identity when sending a message, such that the recipient can be certain that it originated from whom it says it does and accurately conveys what they said. This is achieved using *digital signatures*.

The key point from the security perspective is that the private key cannot be efficiently inferred from the public key. So although everyone has access to Alice's public key, no one is able to counterfeit messages because they cannot create encrypted signatures without access to her private key – signatures can be easily verified but not created.

Because this protocol is implemented using ordinary RSA, albeit with reversed roles for the key pair, it shares the same security strengths and vulnerabilities as RSA public-key cryptography.

Like RSA cryptography, key servers exist, maintaining databases of people's public keys and their associated identities.

Because RSA-encrypted messages are long, digital signature protocols typically do not sign the full document directly. Instead, they create a message digest of the document using a cryptographic hash function, which is signed using RSA. These hash functions have the property that they cannot be forged or manipulated, providing an accurate summary of a document but with extremely low memory overhead (256 bits is typical).

22.5 Hashing

Hash functions are functions that map a long bit string of arbitrary length to a short, fixed-size bit string with quasi-random behaviour,

$$f_{\text{hash}} : \{0,1\}^n \rightarrow \{0,1\}^m, \tag{22.5}$$

for an n-bit input and m-bit output hash, where n is variable and m is fixed. They are an example of 'one-way functions' that are computationally easy to compute in the forward direction but extremely hard to invert. That is, given a hash, it is computationally unviable to find input strings that map to that value.

Hash functions have broad applicability throughout computer science, but here we are most interested in *cryptographic hash functions* for use in cryptography, which impose strong conditions on the difficulty of inversion and their quasi-random characteristics. Most notable, the desired characteristics of a cryptographic hash function include the following:

- The distribution of hashes ought not exhibit any biases, following a uniform distribution with quasi-random behaviour.
- Changing a single bit in the input string ought to flip approximately half of the bits of the hash on average.
- It is computationally efficient to calculate a hash from an input.

- It is computationally complex to find an input that hashes to a given value (that is, they are one-way or trapdoor functions).
- The hashes of two very similar inputs ought to yield hashes that are very different.
- Two different inputs are extremely unlikely to hash to the same value.

The standard cryptographic hash function with mainstream adoption is the 256-bit Secure Hashing Algorithm (SHA256), which generates 256-bit hashes. The algorithm is extremely efficient to implement digitally and exhibits $O(n)$ runtime for input string length n.

Cryptographically, hash functions are useful for creating message digests, which act as a highly condensed checksum of a document that can be utilised in a digital signature.

Note that because the function in general maps longer strings to shorter ones, there are necessarily *collisions* – multiple inputs for a given output. However, for strong cryptographic hash functions their behaviour is sufficiently random that two distinct messages will almost certainly yield completely different hashes (even if the messages are very similar), making it all but impossible for someone to make the claim that Alice said something she did not. This property is extremely important for the security of digital signatures.

23

Attacks on Classical Cryptography

Having introduced the main classes of classical cryptographic protocols, we now turn our attention to their weaknesses and vulnerabilities, against adversaries with both classical and quantum computational resources.

23.1 Classical Attacks

All known classical attacks against any respected classical cryptosystem involve tremendous computational resources. After all, were this not the case the cryptosystem would be considered weak and would never have become widely adopted in the first place!

Brute-Force

The most obvious approach to cracking a cryptosystem is to systematically try out all possible keys until we find one that correctly decodes the encrypted message. This is also the most naïve approach and one that is computationally intractable for real-world key lengths. Specifically, for a key length of k bits ($k = 256$ for AES256), there are 2^k possible keys to try, and on average we will wait for 2^{k-1} trials before choosing the right one. Clearly an average waiting time of 2^{255} is not plausible!

Cryptanalysis

Far better than waiting the age of the universe for the right key to turn up is *cryptanalysis*. Here we study patterns between input and output strings from a cipher utilising a particular key. There are many variations on this but include techniques such as the following [162]:

- Known plaintext attack (KPA): Through alternate means of espionage, the attacker is able to possess *both* a ciphertext and its associated plaintext. Knowing both the input and output to the encryption algorithm may then reveal information about the key relating them. This technique was important to Alan Turing's successful cracking of the German Enigma encryption protocol during World War II.

- Chosen plaintext attack (CPA): The same as a KPA except that the adversary has the ability to choose what the known plaintext is – a more challenging prospect to orchestrate.

- Linear cryptanalysis: A technique for representing ciphers as linear systems, to which KPA are applied.

- Differential cryptanalysis: We analyse how changes in input bits propagate through the cipher to modulate output bits. Typically this type of technique operates as a CPA.

Integer Factorisation

In the case of RSA encryption, whose security derives from the believed computational hardness of factorising large integers, the most efficient known classical algorithm for integer factorisation is the general number field sieve (GNFS), with time complexity

$$O(\exp(O(1)(\log n)^{\frac{1}{3}}(\log\log n)^{\frac{2}{3}})), \qquad (23.1)$$

which scales poorly for large n, keeping in mind that present-day implementations of RSA accommodate key lengths of up to 4,096 bits, as, for example, is implemented by the widely used Pretty Good Privacy (PGP) package.

23.2 Quantum Attacks

Having established that classical attacks against strong classical cryptosystems are quite limited by their implausible computational requirements, what if our adversary now has quantum computational resources? Does this change the game?

Brute-Force

A brute-force attack by a quantum computer does not offer us the exponential improvement attacker Eve might hope for. However, we can gain a quadratic improvement by cleverly exploiting Grover's search algorithm.

To do this, we treat the brute-force cracking algorithm as a satisfiability problem, similar to how Grover's is employed to enhance **NP**-complete problems. Specifically, our oracle implements the code's decryption operation, taking as input a qubit string representing the key. After decoding the message with the key, the oracle runs an appropriate test on the decrypted message to determine whether it is a legitimate decoded message. For example, it could run an English language test – a message decoded incorrectly with the wrong key will appear very random and almost certainly will not pass such a test. The oracle tags an element passing this test, which the Grover algorithm searches for, yielding the associated key.

Note that when performing a brute-force attack against a private encryption key, a quadratic speedup effectively halves the key length in terms of algorithmic runtime, because $O(\sqrt{2^k}) = O(2^{k/2})$. Thus, in the quantum era private key lengths will need to be doubled to maintain an equivalent level of security against brute-force attacks.

This same technique of treating encryption as an oracle within a quantum search algorithm can be utilised to invert hash functions. However, in this case there will necessarily be multiple solutions owing to collisions.

Cryptanalysis

In the case of private-key cryptosystems such as AES, no quantum-enhanced cryptanalytic techniques have been described, which offer an exponential enhancement. Thus, modulo doubling key lengths to counter a Grover attack, these cryptosystems are not regarded as being compromised by quantum computing.

Integer Factorisation

In the case of RSA public-key cryptography the attack is more direct – with access to a scalable quantum computer, Shor's algorithm can be employed to efficiently factorise large integers, allowing private keys to be retrieved from public keys. Unlike the brute-force attacks, which yielded only a quadratic enhancement, Shor's algorithm is exponentially faster than the classical GNFS, requiring runtime of only

$$O((\log n)^2 (\log \log n)(\log \log \log n)). \qquad (23.2)$$

Compare this with the classical case given in Eq. (23.1).

24

Bitcoin and the Blockchain

One of the most exciting new cryptographic applications that has emerged in recent years is the Blockchain, a secure distributed ledger for recording the execution of contracts and transactions. This has enabled cryptocurrencies, most notably Bitcoin, to emerge as a secure digital alternative to conventional fiat currencies.

More recent developments, such as the Ethereum project, develop the distributed ledger further to allow executable code to be committed to the Blockchain, opening the prospects for self-enforcement and -execution of completely arbitrary 'smart contracts', a potential game changer for the operation of financial and derivative markets.

In the Blockchain protocol, the validity of contracts and transactions is recognised collectively by participants using an encrypted digital ledger. The ledger records the complete history of all Blockchain transactions, which are digitally signed by network participants using elliptic-curve public-key cryptography. A democratic process ensures that, provided that a single user does not monopolise the network, recorded transactions are legitimate, recognised collectively and democratically. This is secured by network participants digitally signing off on transactions as they take place.

The Bitcoin protocol builds on top of the Blockchain to create a secure digital cryptocurrency. This requires the introduction of another subprotocol, *mining*, where units of currency ('coins') are created. The protocol cryptographically ensures that there is an upper bound on the number of coins that can exist, thereby preventing forgery and an inflationary blowout in the money supply.

The mining process is based on the computational hardness of inverting (double) SHA256 hashing. A legitimate Bitcoin is defined by a string with a hash satisfying a thoughtfully chosen constraint, specifically one that hashes to a value within some range

$$\epsilon_{\text{lower}} \leq \text{SHA256}(\text{SHA256}(x_{\text{coin}})) \leq \epsilon_{\text{upper}}. \tag{24.1}$$

This is slightly weaker than inverting hash functions but is nonetheless a task that can only be approached via brute-force hashing in the forward direction. This associates computational complexity with the mining process, and hence computational integrity of the money supply, whilst upper bounding the number of unique coins that can exist. This technique is known as 'proof-of-work', for associating something of value with proof that certain amounts of computation were invested into achieving it.[1] This idea was originally borrowed from the Hashcash protocol, where proof-of-work is employed to associate work (and hence monetary value) with sending emails to eliminate automated spamming bots.

The two key algorithms for Bitcoin and the Blockchain are therefore hashing and public-key digital signatures. Both of these are subject to enhanced quantum attacks.

Inverse hashing does not have any known quantum algorithm with exponential improvement; however, using a Grover search one can achieve a quadratic speedup, using the same idea as for enhancing **NP**-complete problems by treating the hash function as a search oracle. This, however, does not pose a fundamental security concern because it will speed up the Bitcoin mining process but does not circumvent the upper bound on the number of coins that may be in existence. Already classical mining has pushed the Bitcoin money supply close to its asymptotic maximum and there is limited room for additional mining.[2]

Elliptic-curve public-key cryptography, like RSA, has a known efficient quantum attack via Shor's algorithm. In the context of implementing digital signatures, this implies that an adversary could fraudulently sign off on illegitimate transactions, thereby committing falsified contracts to the Blockchain.

A detailed investigation into the vulnerability of the Blockchain to quantum attacks was performed by [2]. However, it is nearly impossible to predict the future rate of growth in quantum computer technology and hence over what kind of timescale the Blockchain will be compromised. But it is certain that a full compromise is inevitable at some point in the future when scalable, universal quantum computing becomes a reality.

To address this security threat, quantum-resistant hashing and public-key cryptographic protocols will need to be developed. In the former case this can easily

---

[1] In future implementations of Blockchain protocols, the proof-of-work required for a given protocol can be arbitrarily manipulated to accommodate technological advances in computational power; for example, via the adoption of quantum computing. The amount of work required to satisfy the constraint grows as we narrow the range $\epsilon_{upper} - \epsilon_{lower}$, providing us with much leverage to manipulate the complexity of the proof-of-work, and hence the rate of growth in the money supply, and, equivalently, the rate of inflation.

[2] Bitcoin mining has gained so much traction and become so competitive that desktop PCs have become uneconomical for mining. Instead, miners are resorting to utilising specialised hardware in the form of CUDA cores, FPGAs and ASICs (or by secretly using the company supercomputer while the boss isn't looking).

be achieved by increasing hash lengths to offset the quadratic enhancement offered by Grover's algorithm. In the latter case this will require post-quantum public-key cryptosystems.

Evidently, the life span of existing Blockchain technologies is limited and in the quantum future post-quantum Blockchain algorithms will be required to ensure the survival of cryptocurrencies.

25

Quantum Cryptography

As quantum physics can compromise some important aspects of classical cryptography, can it perhaps be similarly exploited to make new cryptosystems that are immune even to quantum adversaries? Thankfully, the answer is yes ... at least some of the time.

25.1 Quantum Key Distribution

Aside from quantum computing, a central use for quantum technologies is in cryptography [76]. The demand for secure cryptography is now extremely important in the context of electronic commerce and general security of information transmission in the internet age. Electronic currencies such as Bitcoin depend on cryptographic protocols in order to secure the value of assets, assign ownership certificates and secure the currency against fraud. However, such protocols are based on the computational complexity of certain mathematical problems (i.e., computational security) and are not fundamentally secure in the presence of limitless computational resources or quantum computers. Therefore, using quantum mechanical protocols based on physical principles (i.e., information-theoretic security) rather than computational limitations is favourable for future-proofing ourselves.

Quantum key distribution (QKD) protocols facilitate shared, secret randomness, where any intercept-resend (or man-in-the-middle) attack may be detected and rejected, guaranteed by the laws of quantum physics (specifically the Heisenberg uncertainty principle and no-cloning theorem). This shared, secret randomness may subsequently be employed in a one-time pad cipher, presenting us with true information-theoretic security.

The central notion to QKD protocols, in their numerous manifestations, is that measurement of quantum states invokes a wave-function collapse. When measuring a state in a basis for which that state is not an eigenstate, this necessarily changes the state. QKD relies on this simple result from quantum mechanics to reveal any eavesdropper performing an intercept-resend attack via the changes to transmitted quantum states that this would induce.

QKD is a relatively mature technology with several commercial systems already available off-the-shelf, and initial space-based implementations have been successfully demonstrated.

It is easy to see the utility of quantum networks in enabling commodity deployment of QKD – users desire to communicate photons across long-range ad hoc networks, with low loss and dephasing. A global quantum internet would allow quantum cryptography to truly supersede classical cryptography, bypassing the vulnerabilities faced by classical cryptography in the era of quantum computing.

BB84 Protocol

The first described QKD scheme was the *BB84* [20] protocol, which exploits the fact that states encoded in the $\hat{Z}$-basis but measured in the $\hat{X}$-basis (and vice versa) collapse randomly, yielding completely random measurement outcomes, whereas states measured in the same basis in which they were encoded always correctly communicate a single bit of information.

Implemented photonically, BB84 requires only the transmission of a sequence of single photons, polarisation-encoded with random data.

The BB84 protocol is described in detail in Box 25.1 in the context of polarisation-encoded photons, which is the most natural (but not only) setting for this protocol. An example evolution of the protocol is illustrated in Figure 25.1.

Box 25.1 **BB84 QKD protocol using polarisation-encoded photons. Upon completion of the protocol, Alice and Bob share a random bit string for use in a one-time pad cipher, yielding perfect information-theoretic security.**

```
function BB84():
```

1. Alice chooses a random bit, 0 or 1.
2. Alice randomly chooses a basis, $\hat{X}$ or $\hat{Z}$.
3. Depending on the choice of basis, she encodes her bit into the polarisation of a single photon as

$$|0\rangle_Z \equiv |H\rangle,$$
$$|1\rangle_Z \equiv |V\rangle \qquad (25.1)$$

or

$$|0\rangle_X \equiv \frac{1}{\sqrt{2}}(|H\rangle + |V\rangle),$$

$$|1\rangle_X \equiv \frac{1}{\sqrt{2}}(|H\rangle - |V\rangle). \tag{25.2}$$

4. Encoding into the randomly chosen basis, she transmits the randomly chosen bit to Bob.
5. She does not announce the choice of bit or basis.
6. Bob measures the bit in a randomly chosen basis, $\hat{X}$ or $\hat{Z}$.
7. The above is repeated many times.
8. Upon receipt of all qubits, Alice (publicly) announces the basis used for encoding each bit sent.
9. Qubits where Bob measured in the opposite basis to which Alice encoded are discarded, because they will be decorrelated from Alice.
10. The remaining measurement outcomes are guaranteed to yield identical bits between Alice and Bob.
11. Remaining is roughly half as many bits as were sent, which are random but guaranteed to be identical between Alice and Bob.
12. Alice and Bob sacrifice some of their bits by publicly communicating them to check for consistency. This rules out intercept-resend attacks.
13. Privacy amplification may be used to distill the partially compromised key into a shorter but more secret one.

To understand the secrecy of the protocol as described in Box 25.1, suppose that an eavesdropper, Eve, were to perform an intercept-resend attack on the channel between Alice and Bob. At that stage in the protocol Alice had not yet announced her choice of encoding bases, and Eve will not know the bases in which to measure states without randomly collapsing them onto values inconsistent with Alice's encoding. Thus, by sacrificing some of their shared bits, via openly communicating them to one another for comparison, such an attack will be detected with asymptotically high probability. Now Alice and Bob have great confidence that they have a shared, secret, random bit string, which may subsequently be employed in a one-time pad with perfect secrecy.

The BB84 protocol has no measurement timing, mode matching or interferometric stability requirements, making it a very robust protocol, readily achievable

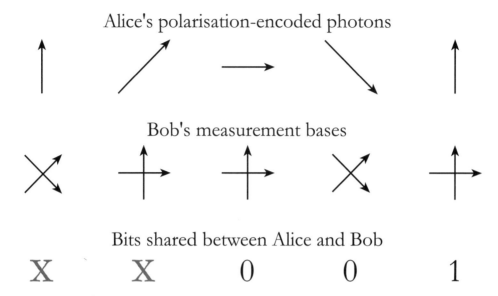

Figure 25.1 Example execution of the BB84 protocol for securely sharing random bit strings between Alice and Bob as per Box 25.1. At the conclusion of the protocol, some bits are discarded (marked 'X'), with those remaining guaranteed to be secret between the two parties.

with present-day photonics technology. The scheme has been adapted to physical architectures beyond just polarisation-encoded photons, such as contiuous-variable encodings.

Privacy Amplification

When Alice and Bob sacrifice and compare a randomly chosen subset of their key bits to detect eavesdroppers, they also need to accept the inescapable fact that their qubits propagated through imperfect channels and were subject to noise en route. This has the same effect as an eavesdropper – it corrupts some of the bits – and it is impossible to distinguish which took place, a noisy channel (which is okay) or an eavesdropper (which is not).

Because the channel was necessarily noisy, Alice and Bob *must* tolerate some number of corrupted bits. But if the corruption came from Eve rather than the noisy channel, they would effectively be tolerating her knowing some of the key. We do not want her to know *any* of the key!

Privacy amplification is a mathematical technique based on hashing algorithms for taking a shared key with a number of unknown compromised bits and distilling it to a shorter key of which Eve has almost zero knowledge.

Specifically, if we know that Eve has compromised t of our n shared random bits, privacy amplification allows us to distill a new key from the compromised one of approximately length $n - t$ over which Eve knows almost nothing.

This is an information-theoretic security result, not a computational security one, thereby rescuing the perfect security of the BB84 QKD protocol.

E91 Protocol

E91 is slightly different to BB84. Here Alice and Bob share an entangled Bell pair provided by a central authority. Then both Alice *and* Bob measure their qubits in random bases. As with BB84, after measuring all qubits, they compare their choices of random bases. When they coincide, they have a shared, random bit. When they do not, they discard their result. From here the remainder of the protocol is the same as for BB84.

Like BB84, E91 has no mode matching or interferometric stability requirements, and Alice and Bob both only require single-photon detection. Unlike BB84, however, E91 requires a central authority that is able to prepare entanglement on demand as a resource.

An advantage of E91 over BB84 is that it does not require a direct quantum communications link between Alice and Bob. The protocol could be mediated from above by a Bell pair-producing satellite within line-of-sight of both Alice and Bob.

Security

Importantly, unlike classical cryptographic protocols, QKD makes no assumptions about the computational complexity of inverting encoding algorithms or trapdoor functions. The protocols are information-theoretically secure and therefore no physically realisable computer – even a quantum computer – can compromise them. Thus, usual cryptanalytic techniques, like linear and differential cryptanalysis [162] or the ability to factor large numbers, that are employed to attack other encryption protocols do not compromise QKD.

However, this is not to say that QKD is actually perfectly secure in real life. Recent history has demonstrated that this is certainly not the case, with many attacks against various quantum cryptographic protocols being described and successfully demonstrated. The reason for this schism between theory and experiment is that no experiment ever *perfectly* mimics the theoretical proposal it is trying to implement. Laboratory components might be imprecise in an unfortunate way, opening up avenues for attack, or they might perform unwanted additional actions that leak information to Eve. The prospects for such so-called side-channel attacks must be carefully considered and satisfactorily addressed.

The best known attack against photonically implemented BB84 is the 'photon-number splitting attack'. This attack targets implementations where Alice's photon source does not produce perfect single-photon states but may have some amplitude of higher photon number. Weak coherent states or spontaneous parametric down-conversion (SPDC) states exhibit this property. The attack is very simple. Eve simply performs a man-in-the-middle attack but not of an intercept-resend variety. Rather than intercepting the entire channel, she inserts a low-reflectivity beam splitter and measures only the reflected mode; the other follows its desired trajectory to Bob. Now there is a chance that Eve can extract just one of the multiple photons in the signal, such that Bob still receives a photon. Eve holds the split-off signal in memory until the classical communication of encoding bases, at which point she measures all of her split signals in the correct basis, thereby recovering the associated secret key bit.

This trivial attack vector clearly demonstrates the importance of well-considered engineering decisions when physically implementing QKD. No piece of hardware is ever 100% to specification!

Public-Key Cryptography

The BB84 protocol is used exclusively for private-key cryptography. For many applications (notably digital signatures and easy key exchange with unidentified parties), public-key cryptosystems would be highly desirable.

Are there any viable public-key quantum protocols that could fill the vacancy of the soon-to-be-compromised RSA? Unfortunately, the answer is 'not yet'. As appealing as it would be, and despite many highly intelligent people putting their minds to it, to date no one has presented a viable public-key quantum cryptosystem.

This is problematic because when quantum computing becomes a reality it will immediately compromise the classical public-key cryptosystems we all rely on on a daily basis, and it would be highly desirable for a quantum replacement to be available to fill its shoes.

25.2 Hybrid Quantum/Classical Cryptography

As discussed, the RSA public-key cryptosystem is vulnerable to an efficient quantum attack, whereas private-key schemes like AES are not (believed to be). Thus, combining QKD schemes with private-key classical schemes does not compromise security in the quantum era.

Why would we combine quantum and classical encryption techniques when QKD is already provably secure, whereas the classical schemes are not?

In the near future, as QKD schemes begin their rollout in space and on Earth, random bits from the QKD implementation will be very expensive and exhibit low bandwidth. Suppose that we wanted to securely videoconference across the globe. For just a single user this would require megabits per second of shared random bits, which will quickly saturate the capacity of overhead quantum satellites.

Instead, let us use the QKD system to securely share just a 256-bit private session key between two users. This is subsequently employed for AES256 encryption that operates entirely over the classical network, which we regard as extremely cheap and high-bandwidth. Importantly, unlike one-time pad implementations, this session key may be reused. Now we have a hybrid system that is not quantum-compromised but that overcomes the cost and bandwidth issues associated with emerging QKD networks.

Though such a hybrid scheme is not information-theoretically secure (AES is not proven to be quantum safe), the computational security assumptions are far stronger than for, say, RSA, because there are no known efficient quantum attacks against strong private-key schemes.

25.3 Quantum Anonymous Broadcasting

The previously described protocols all focussed on preserving the secrecy of messages. Alternately, it may not be the message that is sensitive but rather the identity of the person who says it. *Anonymous broadcasting* is a protocol for achieving this.

Consider the following scenario. A group of users share a classical broadcast channel that anyone is able to transmit to and everybody is able to listen to unencrypted. But it is of importance that the identity of whoever broadcasts to the channel be kept secret from all users. A scheme exists for achieving this quantum mechanically using shared Greenberger-Horne-Zeilinger (GHZ) states – *quantum anonymous broadcasting* (QAB).

Let there be a (trusted[1]) server that distributes GHZ states (of arbitrary numbers of qubits) to a group of users, one qubit per user. This can be prepared as described in Section 13.4. Now if every user measures in the $|\pm\rangle = \frac{1}{\sqrt{2}}(|0\rangle + |1\rangle)$ basis the joint *parity* (i.e., whether an even or odd number of +'s were measured) is guaranteed to be even. For example, all users might measure $|+\rangle\langle+|$, or exactly 2, but never exactly 1 or 3.

On the other hand, if a $\hat{Z}$ gate were applied to any one qubit, this would flip the parity outcome. Note that a GHZ transforms according to

---

[1] Note that if the server is not trusted, he could easily conspire to reveal people's identities by distributing $|+\rangle^{\otimes n}$ states instead of GHZ states.

$$\hat{Z}_i \frac{1}{\sqrt{2}}(|0\rangle^{\otimes n} + |1\rangle^{\otimes n}) \rightarrow \frac{1}{\sqrt{2}}(|0\rangle^{\otimes n} - |1\rangle^{\otimes n}) \; \forall\, i \qquad (25.3)$$

for any qubit i. This invariance in the location of the $\hat{Z}$ gate is the basis for the anonymity of the protocol. If a user wishes to broadcast '0' he does nothing, whereas if he wishes to broadcast '1' he applies a $\hat{Z}$ gate to his local qubit.

Finally, all users measure their qubits in the $\pm$-basis and publicly (without encryption) broadcast their measurement outcomes. All users now see all other users' measurement outcomes and are able to calculate the collective parity of the measurements. Now if the parity is even, the speaker must have said '0', whereas if it is odd he must have said '1'. The protocol is shown in Figure 25.2.

Note that the scheme can be slightly simplified: rather than the speaker applying the $\hat{Z}$ to his qubit, upon announcing his measurement outcome he simply lies about his outcome and flips it. This follows simply because a $\hat{Z}$ gate prior to a $\pm$ measurement bit-flips the classical measurement outcome, $\hat{Z}|\pm\rangle = |\mp\rangle$.

There are no constraints on time ordering of the measurements, nor, much like BB84, are there any interferometric stability requirements (not including the GHZ preparation stage), making this protocol very experimentally practical and robust over long distances.

Because of the time invariance in the measurements, distribution and measurement of the GHZ states can be performed well in advance of the actual message

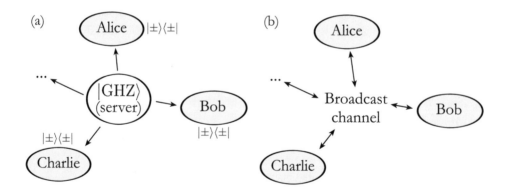

Figure 25.2 Protocol for quantum anonymous broadcasting. (a) A central trusted server prepares GHZ states and distributes them amongst a group of users, one qubit per user. All users measure in the $\pm$-basis. (b) All users classically broadcast their measurement outcomes yielding shared random parity. During broadcast, the broadcaster lies about his measurement outcome to flip the joint parity if he wishes to transmit '1', or tells the truth to transmit '0'. The joint parity encodes the message of the anonymous user, which all listeners are able to recover. Importantly, only one user may broadcast at a time, otherwise the recovered message will be given by the XOR of all the simultaneously broadcast messages.

broadcast. This allows us to treat 'shared parity' as a fundamental resource for the QAB cryptoprotocol.

Because the parity sharing can be isolated from the broadcasting stage, it is unimportant if the GHZ source is nondeterministic or the channels for distributing it lossy. We can instead simply repeat GHZ distribution over and over at high repetition rate, post-selecting upon measurement outcomes where all users signal that they successfully received and measured their photons.

For these reasons, this scheme lends itself readily to photonic implementation, provided a reliable GHZ preparation circuit. The scheme has since been ported to operate on distributed toric codes to facilitate error correction of the distributed GHZ states [111].

25.4 Quantum Voting

The security of voting systems and the anonymity of their voters are pressing issues in the modern free world, and there have been countless high-profile instances of voting systems being compromised nefariously.

Based on ideas similar to quantum anonymous broadcasting is quantum voting, whereby a group of parties can anonymously vote such that no party, including the tallyman, is able to learn any individual voter's vote but at the conclusion all are able to see the collective outcome of the vote.

There are a multitude of different models for voting, and a number of quantum implementations for them have been described. The two most well-known are the following:

- Binary voting: whereby each party votes 'yes' or 'no'.
- Anonymous surveys: whereby each party votes a number and we wish to determine the sum of all of the votes.

Conceptually, the scheme is similar to quantum anonymous broadcasting in that it hides votes in phases within an entangled state that are not accessible to individual parties but are rather a global property of the state. The scheme relies on the preparation and distribution of a particular entangled state as a resource. Unfortunately, this particular state is not one that is known how to be trivially prepared optically and would therefore lend itself well to the outsourcing of preparation and distribution to a capable host via the quantum internet.

26

Attacks on Quantum Cryptography

For a quantum key distribution (QKD) system, information-theoretic security is achieved only when security against collective, coherent attacks is proven. We must not make assumptions about the limitations of our adversaries. For a more comprehensive review on quantum cryptography, see [132].

Hacking attacks in this context exploit weaknesses in the physical implementation, rather than weakness of the theory – so-called side-channel attacks. Some examples of weaknesses that allow zero-error attacks include the following:

- Losses: Genuinely lossy channels or components are indistinguishable from ideal ones where some of the signal has been tapped off by Eve. Therefore, we must always assume the worst: that whatever is lost from our system is in the hands of Eve.
- Imperfect components: Our physical implementation might just not be operating strictly according to the theory.
- Correlations: Mutual information between signals in our system may leak information from the secure system to the environment, where Eve might be waiting patiently.

When considering the security of noisy channels, one must assume that all noise is due to manipulation by an eavesdropper – the worst-case scenario. An attempted attack is considered successful if it can be proven that the eavesdropper can gain a nonnegligible (i.e., not exponentially small) amount of mutual information with the final secret key established between Alice and Bob, without alerting them. Some of the best-known attacks follow.

26.1 Beam Splitter and Photon Number–Splitting Attacks

The security proofs for many discrete variable protocols assumes that Alice's signals consist of single photons. However, true single-photon soures are not yet widely available, and QKD systems often make use of strongly attenuated laser pulses, and there is some probability of the source emitting multiple photons. This fact can be exploited by Eve, who employs the photon number splitting (PNS) attack [18]: Eve can perform a nondemolition measurement to determine the number of photons in the signal; she steals the excess photons and sends the rest to Bob. She stores these in her quantum memory and waits until the classical communication between Alice and Bob, hence finding out Alice's preparation basis.

A beam splitting (BS) attack translates the fact that any signal lost over a channel is acquired by Eve. Here, Eve induces losses in the communications channel by putting a beam splitter outside Alice's device and then forwards the remaining photons to Bob. The BS attack does not modify the optical mode that Bob receives: it is therefore always possible for lossy channels and does not introduce any errors.

A method used to counter the PNS attack is the decoy-state method [89, 110]. In the decoy-state protocol, Alice randomly replaces some of her signal states with multiphoton pulses from a decoy source. Eve cannot distinguish between decoy pulses from the encoding signals and can only act identically on both. In the post-process stage, Alice public announces which states were the decoy pulses. The trusted parties can then characterise the action of the channel on the multiple-photon pulses and detect the presence of a PNS attack.

26.2 Trojan Horse and Flashback Attacks

Another family of hacking that can be used against discrete-variable QKD is Trojan horse attacks. These attacks involve Eve probing the settings of Alice and Bob by sending light into their devices and collecting the reflected signal. The first of this kind of attack actually came for free for the eavesdropper [160]: it was discovered that some photon counters emit light when a photon is detected [104]. If the emitted light carries correlated information about which detector was triggered, it must be prevented from leaking outside the secure space and becoming accessible to Eve.

In general, Eve probes into the optical channel that Alice and Bob use to communicate. She sends her own states into Alice's system, which will reflect off the same apparatus Alice uses used to encode her signal. Eve's states can be imprinted upon some information about the encoding used by Alice, when Eve measures them. She can then use the result of this measurement, combined with some operation on Alice's signals, to make a best estimate of the quantum state that Alice sent to Bob, thus giving her some nonnegligible mutual information with the key [183].

26.3 Detector Attacks

The faked-state attack is based on the weak-laser implementation of BB84. Here, Eve manipulates Bob's detectors to force him to measure in the same basis.

It exploits the fact that the detectors may have a dead time, and the eavesdropper can trigger the detector whenever she chooses. It follows that Bob's detection outcomes are controlled by the eavesdropper.

Eve can also go beyond detector blinding. She can send in a powerful laser pulse to optically damage components in the QKD system and permanently change its characteristics [91]. If the new characteristics then assist the eavesdropper in an attack without Alice or Bob being notified, the security of the QKD system would be severely compromised.

A more detailed discussion on attacks on physical implementation can be found in [91].

Part VI

Quantum Computing

Because quantum computing is perhaps the most exciting of the emerging quantum technologies, which we treat as the foremost application for the quantum internet, we now introduce quantum computing, covering models and physical implementations for realising it and some of its well-known algorithmic applications.

27

Models for Quantum Computation

We begin by reviewing the models for quantum computation that we will refer to throughout this work. There are various approaches to implementing and representing quantum computations. We now briefly introduce the ones most relevant to our discussions on networked quantum computation.

27.1 Circuit Model

The *circuit model* is the conventional and most intuitive approach for expressing quantum algorithms, decomposing them into chronological sequences of elementary operations, comprising state preparation, single- and multi-qubit gates, measurement and classical feedforward. We recommend referring to the introductory sections of [127] for a far more comprehensive introduction to quantum circuits than is presented here. This model will be naturally intuitive to those familiar with classical circuit diagrams, albeit with some important differences, such as time ordering.

Figure 27.1 illustrates a simple three-qubit quantum circuit comprising all of these elements. The interpretation of this diagram is as follows:

- Horizontal lines represent individual qubits.
- Time flows from left to right (feedback is not allowed in the typical formalism for this representation).
- The three input qubits are labelled on the far left as $|\psi_1\rangle$, $|\psi_2\rangle$ and $|\psi_3\rangle$.
- Single-qubit gates are denoted as boxes containing the name of the associated unitary operation. Here, the examples are the Hadamard ($\hat{H}$), Pauli bit-flip ($\hat{X}$), Pauli bit-phase-flip ($\hat{Y}$) and Pauli phase-flip ($\hat{Z}$) gates,

$$\hat{H} = \frac{1}{\sqrt{2}} \begin{pmatrix} 1 & 1 \\ 1 & -1 \end{pmatrix},$$

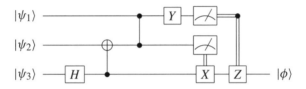

Figure 27.1 Simple example of a quantum circuit on three qubits, comprising several single- and two-qubit quantum gates and measurements. Rows represent qubits, and time flows from left to right.

$$\hat{X} = \begin{pmatrix} 0 & 1 \\ 1 & 0 \end{pmatrix},$$

$$\hat{Y} = \begin{pmatrix} 0 & -i \\ i & 0 \end{pmatrix},$$

$$\hat{Z} = \begin{pmatrix} 1 & 0 \\ 0 & -1 \end{pmatrix}. \tag{27.1}$$

- Two-qubit gates are denoted by vertical lines between the respective qubits.
- The maximally entangling two-qubit controlled-NOT (CNOT) gate is denoted via a control (•) and a target (⊕),

$$\text{CN̂OT} = \begin{pmatrix} 1 & 0 & 0 & 0 \\ 0 & 1 & 0 & 0 \\ 0 & 0 & 0 & 1 \\ 0 & 0 & 1 & 0 \end{pmatrix}. \tag{27.2}$$

This is the quantum equivalent of the classical XOR gate, flipping the target ($\hat{X}$) if the control is on.

- All quantum gates have the same number of input as output qubits. This is a necessary condition for the unitarity of quantum gates ($\hat{U}^\dagger \hat{U} = \hat{I}$).
- The maximally entangling two-qubit controlled-phase (CZ) gate is denoted by two targets (•) (the gate operates symmetrically on its two qubits),

$$\hat{\text{CZ}} = \begin{pmatrix} 1 & 0 & 0 & 0 \\ 0 & 1 & 0 & 0 \\ 0 & 0 & 1 & 0 \\ 0 & 0 & 0 & -1 \end{pmatrix}, \tag{27.3}$$

applying a phase-gate ($\hat{Z}$) to the target if the control is on.

- The 'meter' symbol represents a classical measurement in the Pauli $\hat{Z}$-basis (the computational or logical basis).

- Double lines represent classical feedforward of measurement outcomes, controlling a subsequent gate.

The circuit in Figure 27.1 can be interpreted mathematically as implementing the following operation:

$$|\phi\rangle = \hat{Z}_3^{m_1} \cdot \hat{X}_3^{m_2} \cdot \hat{M}_2 \cdot \hat{M}_1 \cdot \hat{Y}_1$$
$$\cdot \hat{CZ}_{1,2} \cdot \hat{CNOT}_{3,2} \cdot \hat{H}_3 \cdot |\psi_1\rangle \otimes |\psi_2\rangle \otimes |\psi_3\rangle, \qquad (27.4)$$

where m_1 and m_2 are the binary measurement outcomes of the two single-qubit $\hat{Z}$-basis measurements, $\hat{M}_1$ and $\hat{M}_2$.

Using the circuit model, arbitrary quantum computations can be elegantly and intuitively represented. To enable *universal* quantum computation within this model, a *universal gate set* must be available at our disposal. Most commonly, this is chosen to be the maximally entangling two-qubit CZ or CNOT operation, in addition to arbitrary single-qubit gates. Any quantum (i.e., **BQP**) algorithm may be efficiently decomposed into a polynomial-depth circuit comprising elements from this universal gate set. Note that the universal gate set is not unique, and there are many distinct sets. However, this set must contain at least one entangling operation acting on two or more qubits (such as a CZ or CNOT gate) and at least one non-Clifford gate.[1]

27.2 Cluster States

The *cluster state* model for quantum computation [137, 138, 126] (also referred to as the *one-way*, *measurement-based* or *graph state* models for quantum computation) is an extremely powerful paradigm that warrants treatment of its own, owing to its significant distinction from the more familiar circuit model and its applicability to distributed models for quantum computation, to be discussed in Section 29.2.

In the cluster state model, we begin by preparing a particular, highly entangled state, called a *cluster state* or *graph state*. The state is associated with a graph G, comprising vertices, V, and edges, E,

$$G = (V, E), \qquad (27.5)$$

of some topology, although rectangular lattice graphs are usually considered because they are sufficient for universal quantum computation.[2] That is, they act as a 'substrate' for implementing arbitrary quantum computations.

---

[1] The Clifford group is that which commutes with the CNOT gate, such as the Pauli group.
[2] Note that the graph upon which a cluster state resides is not to be confused with the network graph. Rather, it is just a convenient graphical representation for a class of multi-qubit states.

In the graph, vertices represent qubits initialised into the

$$|+\rangle = \frac{1}{\sqrt{2}}(|0\rangle + |1\rangle) \tag{27.6}$$

state, and edges represent the application of maximally entangling CZ gates between vertices,

$$|\psi\rangle_{\text{cluster}} = \prod_{e \in E} \hat{CZ}_e \cdot \bigotimes_{v \in V} |+\rangle_v. \tag{27.7}$$

Alternatively, but equivalently, cluster states may be defined in the stabiliser formalism. Specifically, a cluster state is defined to be the joint $+1$ eigenstate of all of the stabilisers,

$$\hat{S}_v = \hat{X}_v \prod_{i \in n_v} \hat{Z}_i, \tag{27.8}$$

where there is one stabiliser $\hat{S}_v$ per vertex v, and n_v is the set of vertices neighbouring v. The cluster state therefore satisfies

$$\hat{S}_v |\psi\rangle_{\text{cluster}} = |\psi\rangle_{\text{cluster}} \,\forall\, v, \tag{27.9}$$

and the full set of stabilisers, $\hat{S}_v$, over all vertices v is sufficient to fully characterise the cluster state, $|\psi\rangle_{\text{cluster}}$, for a given graph topology.

An example of a rectangular lattice cluster state is presented in Figure 27.2. Cluster states are easily encoded optically using photonic polarisation encoding and therefore readily lend themselves to optical networking.

Having prepared this state, the computation is implemented purely via a well-orchestrated routine of single-qubit measurements. The order and basis in which they are performed (which depends on previous measurement outcomes in general; i.e., we require fast-feedforward) then stipulates the computation. In the context of distributed computation, this requires classical communication between nodes.

Mapping a circuit model computation to a cluster state topology can be most naïvely performed by taking a circuit acting on n_{qubits} qubits with depth n_{depth}, preparing an $n_{\text{qubits}} \times n_{\text{depth}}$ rectangular lattice cluster and 'etching' the circuit directly into the cluster state substrate. To perform this mapping we choose a universal gate set comprising CZ and single-qubit gates, retaining vertical edges where CZ gates ought to be present and eliminating the remaining vertical edges. Now we have a substrate that looks topologically very much like its equivalent circuit construction and the computation proceeds chronologically in the same manner. The only conceptual distinction is that in the circuit model gates are directly applied chronologically to a set of qubits, whereas in the cluster state model gate teleportation is effectively implemented upon each measurement, with

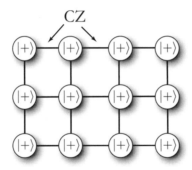

Figure 27.2 Example of a 4×3 rectangular lattice cluster state. Each vertex in the graph represents a qubit initialised into $|+\rangle = \frac{1}{\sqrt{2}}(|0\rangle + |1\rangle)$. Edges represent the application of CZ gates between qubits (CZ gates commute, so the order is unimportant). Of sufficient dimension, states of this topology enable universal measurement-based quantum computation, whereby computation proceeds purely via single-qubit measurements and all entangling operations have been commuted to the state preparation stage. Because CZ gates commute, the preparation of cluster states is time independent and easily implemented in a distributed or parallelised manner. The time ordering of the single-qubit measurements is dependent on the structure of the graph and the algorithm.

the action of gates accumulating as these teleportations are successively applied. A simple example of this notion is shown in Figure 27.3.

The distinctive feature of this model is that all of the entangling CZ gates are performed at the very beginning of the protocol, during the state preparation stage. The algorithm itself is purely measurement based, requiring only single-qubit measurements (no entangling measurements).

An alternate interpretation of the cluster state model is that it is a complicated network of state and gate teleportation protocols. Specifically, a CZ gate with a $|+\rangle$ state as a resource, followed by measurement of one of the two qubits, acts as a single-qubit teleporter, as shown in Figure 27.4.[3] Thus, with a substrate state of CZ gates applied between $|+\rangle$ states, the single-qubit measurements progressively teleport the input state through the graph topology, at each stage accumulating the action of more gates, which are related to the choices of the previous single-qubit measurement bases and the graph topology.

The cluster state formalism has proven very useful, enabling the development of models for linear optics quantum computing, orders of magnitude more efficient than the originally proposed protocol. It has been found that bonding strategies – i.e., the order in which smaller clusters are fused into larger ones when using

---

[3] This is an alternative but equivalent implementation for quantum state teleportation to that presented in Section 16.3.

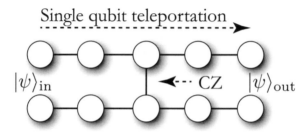

Figure 27.3 Simple example of a cluster state that performs a computation comprising single-qubit operations and a CZ gate between two logical qubits. Let the two horizontal chains represent our two logical qubits. After inputting our input state from the left, we progressively measure out the cluster state qubits chronologically from left to right. Upon each single-qubit measurement, the choice of measurement basis teleports the action of a single-qubit gate. These accumulate sequentially. When we reach the point of measuring the two cluster state qubits joined with the vertical edge, the logical qubits accumulate the action of a CZ gate between them, because this is identically what that vertical edge physically corresponds to. Reaching the final two qubits, one from the upper rail and one from the lower, we obtain our two output logical qubits.

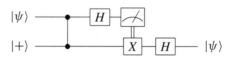

Figure 27.4 The single-qubit teleporter, based on a CZ gate, a single-qubit measurement and classical feedforward.

nondeterministic gates – plays a major role in resource overhead, and much work has been performed on efficient preparation strategies for various topologies [125, 13, 37, 17, 82, 145, 96, 98, 145, 97, 42, 41].

These cluster states are highly valuable, given their computational power and the ability to communicate them from Alice, who is able to prepare them, to Bob, who lacks the technology, would be a boon for Bob.

It would be most practical, economical and resource efficient to have a single, well-equipped server with the ability to prepare such states, who does so on behalf of everyone else and communicates the fresh cluster states to them over the quantum internet (for a price, perhaps).

Importantly, the preparation of cluster states is readily parallelised. All of the entangling CZ operations commute, the order in which they are applied is irrelevant and a rectangular lattice cluster is completely uniform. Thus, the graphs representing smaller cluster states may be easily 'fused' together to form larger cluster states using, for example, CZ gates. Several other types of entangling gates can also be employed, such as polarising beam splitters – so-called *fusion gates* [37]. This allows the preparation of cluster states to be performed in a 'patchwork quilt'–like manner: a number of nodes each prepare small lattice clusters, which are all put side by side and stitched together using CZ gates. This type of distributed state preparation is a perfect application for in-parallel distributed quantum processing.

Consider the scenario whereby Alice requests a large cluster state from Bob but, though she was unable to prepare the cluster state herself, she has the technological ability to perform the measurement-based computation on the state (i.e., simple single-qubit measurements). This would effectively bypass the need for secure quantum computation on Bob's hardware altogether, enabling computation with *perfect* secrecy, because no foreign parties would be involved in the computation stage and no secret data is communicated; only the *substrate* for the computation is communicated, which could be used for any purpose whatsoever. By commuting all of the technologically challenging aspects of a quantum computation to the state preparation stage, we can effectively mitigate the need for blind quantum computing entirely, because the 'hard work' has been done in advance by the host and Alice gets to fulfil the computation on her own, completely bypassing poor old Bob, who was just dying to read Alice's secret love letters before processing them into Hallmark cards.

There are several cluster state identities we will utilise later, summarised in Figure 27.5.

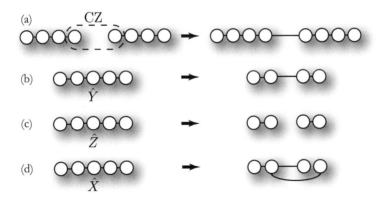

Figure 27.5 Several cluster state identities, demonstrated in the case of linear clusters. (a) A CZ gate between two qubits creates an edge between them in the graph. (b) Measurement of a qubit in the Pauli $\hat{Y}$ basis removes that qubit from the graph, whilst creating new edges between the neighbouring qubits. (c) Measurement of a qubit in the Pauli $\hat{Z}$ basis removes that qubit and any neighbouring edges. (d) Measurement of a qubit in the Pauli $\hat{X}$ basis removes that qubit, leaving one of its neighbours as a 'dangling node'.

When using nondeterministic gates (i.e., ones that probabilistically fail) to prepare cluster states, there are approaches to nonetheless preparing ideal cluster states. There are two main approaches that have become particularly well known.

The first is to use the ideas of *microclusters* and cluster state recycling to incrementally build up larger clusters, progressing as a random walk, which is biased in the direction of state growth. This approach is discussed in more detail in Section 29.4.

The second approach is to borrow techniques from percolation theory to simply tolerate defects in a cluster state lattice by working around them. Specifically, if the defect probability (i.e., probability of a missing vertex or edge) is below some *percolation threshold*, $p_{\text{defect}} \leq \epsilon_{\text{threshold}}$, in the asymptotic limit we are guaranteed that routes exist through the lattice, enabling the required flow of information. This allows defective graphs to be employed for quantum computation.

27.3 Restricted Models for Quantum Computation

In the near future we are unlikely to have devices with the full power and versatility of universal quantum computers. Instead, we will gradually evolve towards that challenging goal via many incremental, intermediate steps. Those steps will take us along a path of restricted quantum computers that solve specific problems of relatively small size and are not fault-tolerant. We can probably expect them to contain on the order of hundreds of qubits in the coming few years, as of the time of writing this book.

One thing is for certain: full-fledged quantum computing will remain an extremely ambitious goal for some time, but we will learn a lot from traversing the path towards it and, it his hoped, uncover new restricted quantum applications along the way.

27.4 Fault Tolerance

The purpose of fault tolerance is to error-protect an entire computation and all of the information residing within it at every stage throughout its execution.

This is achieved by extending techniques from quantum error correction (QEC) to achieve *fault-tolerant quantum computing*. The primary difficulty here is that a quantum computation is not a passive operation but involves the successive application of a potentially enormous number of quantum gates, each of which is subject to its own error processes, all of which must be mitigated for the computation to succeed.

Because a quantum computation is not a passive operation but highly active, fault-tolerance protocols are also active and it does not suffice to simply perform an encoding at the beginning and error correction at the end. Instead, error correction procedures must be applied repeatedly throughout execution, at each stage projecting the encoded computation onto an error-free state.

The concept of fault tolerance in computation is not new; it was first developed in relation to classical computing [184, 71, 8]. However, in recent years the precise manufacturing of digital circuitry has made large-scale error correction and fault-tolerant circuits largely unnecessary.

The basic principle of fault tolerance is that the circuits used for gate operations and error correction procedures should not cause errors to cascade. Quantum gates not only covert errors; i.e., a Hadamard operation can convert an X-error into a Z-error and visa versa but multi-qubit gates can also *copy* errors. Hence, if quantum circuits are not designed carefully, a correctable number of *physical* errors could occur that are consequently copied so many times that they overwhelm the error correction capabilities of the encoding scheme.

Fault tolerance, in the context of error correction, is a function of how circuits and protocols are implemented, not a function of the underlying physical hardware. It is assumed that all single-qubit gates can introduce single-qubit errors at some probability, p, and it is assumed that all two-qubit gates will *copy* pre-existing errors that exist at the input and also have the possibility of introducing a two-qubit correlated error on the two qubits, with probability p.

In some cases there are examples of higher order gates being defined as primitives; for example, the three-qubit Toffoli gate. However, it should be noted that in almost all cases, the physical implementation of these multi-qubit gates occurs through an implicit decomposition into single- and two-qubit gates. This is due to the fact that in the vast majority of physical systems, the highest order coupling term in a system Hamiltonian is weight 2. Higher weight coupling terms, which would be required to enable native multi-qubit gates (i.e., weight 3 terms in the Hamiltonian would be needed to natively enact a Toffoli gate), simply do not arise in natural and easily controllable quantum systems.

To determine how errors are copied by gate operations, an error operator E is conjugated through the gate operation to create a new error operator, $E' = G^\dagger E G$, for some gate unitary, G. A single-qubit example is the transformation of X-errors to Z-errors and visa versa through a Hadamard gate, due to the identity $\hat{X} = \hat{H}\hat{Z}\hat{H}$ and $\hat{Z} = \hat{H}\hat{X}\hat{H}$.

A two-qubit example is more involved because we need to check all combinations of error mappings on both qubits involved in the gate. If $\hat{G} = \text{CNOT}$, we can examine how X- and Z-errors change via G,

$$\text{CNOT}(I \otimes X)\text{CNOT} = I \otimes X$$
$$\text{CNOT}(X \otimes I)\text{CNOT} = X \otimes X$$

$$\text{CNOT}(I \otimes Z)\text{CNOT} = Z \otimes Z$$
$$\text{CNOT}(Z \otimes I)\text{CNOT} = Z \otimes I, \tag{27.10}$$

where the notation $A \otimes B$ are error operators, $\{A, B\} \in \{I, X, Y, Z\}$, on qubits 1 and 2 of the gate and $\hat{G} = \hat{G}^\dagger = \text{CN}\hat{\text{O}}\text{T}$.

So, for a controlled-NOT operation, X-errors are copied from control qubit to target and Z-errors are copied from target to control. Pre-existing X-errors on the target qubit or Z-errors on the control qubit are unchanged through the gate.

The fact that quantum circuits can cause errors to be copied implies that if circuits are designed badly, errors can cascade during error correction protocols even when only one or two *physical* errors actually took place. Error correction codes have a finite correcting power; i.e., the Steane code will deterministically correct an arbitrary *single* qubit error, but if more than a single error occurs between correction cycles, logical errors are likely to be induced.

Fault tolerance is a discrete feature of a quantum circuit construction; either a construction is fault tolerant or it is not. However, what is defined to be fault tolerant can be a function of what type of error correction code is used. For example, for a single error correcting code, fault tolerance is defined as follows:

- A single error will cause *at most* one error in the output for each logical qubit block.

However, if the quantum code employed is able to correct multiple errors, then the definition of fault tolerance can be relaxed; i.e., if the code can correct three errors, then circuits may be designed such that a single failure results in at most two errors in the output (which is then correctable). In general, for a code correcting $t = \lfloor (d-1)/2 \rfloor$ errors, fault tolerance requires that $\leq t$ errors during an operation does not result in $> t$ errors in the output for each logical qubit.

The Threshold Theorem

The threshold theorem is a consequence of fault-tolerant circuit design and the ability to perform dynamical error correction. Rather than present a detailed derivation of the theorem for a variety of noise models, we will instead take a very simple case where we utilise a quantum code that can only correct for a single error, using a model that assumes uncorrelated errors on individual qubits. For more rigorous derivations of the theorem see [3, 79, 3].

Consider a quantum computer where each physical qubit experiences either an X and/or Z error independently with probability p per gate operation. Furthermore, it is assumed that each logical gate operation and error correction circuit is designed fault-tolerantly and that a cycle of error correction is performed after each elementary *logical* gate operation. If an error occurs during a logical gate operation, then the fault-tolerant constructions ensure that this error will only propagate to at most one error in each block, after which a cycle of error correction will remove the error.

Hence, if the failure probability of unencoded qubits per time step is p, then a single level of error correction will ensure that the logical step fails only when two (or more) errors occur. Hence, the failure rate of each logical operation, to leading order, is now $p_L^1 = cp^2$, where p_L^1 is the failure rate (per logical gate operation) of a first-level logical qubit and c is the upper bound for the number of possible two-error combinations that can occur at a physical level within the circuit consisting of the correction cycle + gate operation + correction cycle.

We now repeat the process, re-encoding the computer such that a level 2 logical qubit is formed, using the same $[[n, k, d]]$ quantum code, from n level 1 encoded qubits. It is assumed that all error correcting procedures and gate operations at the second level are self-similar to the level 1 operations (i.e., the circuit structures for the level 2 encoding are identical to the level 1 encoding). Therefore, if the level 1 failure rate per logical time step is p_L^1, then by the same argument the failure rate of a two-level operation is given by $p_L^2 = c(p_L^1)^2 = c^3 p^4$. This iterative procedure is then repeated (referred to as concatenation) up to the kth level, such that the logical failure rate, per time step, of a k-level encoded qubit is given by

$$p_L^k = \frac{(cp)^{2^k}}{c}. \tag{27.11}$$

Equation (27.11) implies that for a finite *physical* error rate, p, per qubit, per time step, the failure rate of the kth-level encoded qubit can be made arbitrarily small by simply increasing k, dependent on $cp < 1$. This inequality defines the threshold. The physical error rate experienced by each qubit per time step must be $p_{th} < 1/c$ to ensure that multiple levels of error correction reduce the failure rate of logical components.

Hence, provided that sufficient resources are available, an arbitrarily large quantum circuit can be successfully implemented, to arbitrary accuracy, once the physical error rate is below threshold. Initial estimates at the threshold, which gave $p_{th} \approx 10^{-4}$ [101, 3, 79], did not sufficiently model physical systems in an accurate way. Recent results [170, 174, 175, 112, 11] have been estimated for more realistic quantum processor architectures, showing significant differences in threshold when architectural considerations are taken into account. The most promising thresholds that have been calculated for expected circuit-level noise are based on surface codes [188, 187, 69, 169], with thresholds slightly less than 1%. This has now become the target for experimental groups because a large number of scalable systems architectures utilise the surface code as the underlying correction model [75, 86, 106, 124, 93, 117].

28

Quantum Algorithms

The ultimate goal of quantum computing is to implement algorithms with a quantum speedup in comparison to classical algorithms. The degree of speedup achieved varies between algorithms, and it is important to note that not every classical algorithm exhibits any speedup when implemented quantum mechanically.

To provide context for the excitement of quantum computing and motivate interest in their development, we now summarise some of the key quantum algorithms that have been described exhibiting quantum speedup.

28.1 Deutsch-Jozsa

The first quantum algorithm demonstrating a provable improvement over the best classical algorithm was the Deutsch-Jozsa algorithm [52]. Unfortunately, the algorithm solves a very contrived problem, designed for the purposes of demonstrating post-classicality rather than solving a problem of actual practical interest. Nonetheless, the algorithm is straightforward to explain and understand, making it a useful starting point in understanding quantum algorithms and the computational enhancement they may offer.

The algorithm relies on a 'black box', referred to as an *oracle*, which takes an input bit string and outputs a single bit, evaluating the function $f(x)$ for the n-bit input bit string x. In this contrived problem $f(x)$ is guaranteed to be either *uniform* or *balanced*. In the former case, the output to the oracle is always $f(x) = 0$ or always $f(x) = 1$, but it does not matter which – they simply must always be the same. In the latter case, the output is $f(x) = 0$ for exactly half the inputs x and $f(x) = 1$ for the other half of x, but the ordering of which inputs generate which outputs may be arbitrary. The goal of the algorithm is to determine whether $f(x)$ is uniform or balanced using the least number of queries to the oracle.

Though it is clear that the dimensionality of the input state space is exponentially large, 2^n, it is fairly obvious that a trivial **BPP** algorithm exists for solving this problem with confidence exponentially asymptoting to unity against the number of oracle queries. We simply evaluate the oracle for randomly chosen inputs. If we measure any occurrences of measurement outcomes that are not all 0 or all 1 we know with certainty that the function must have been balanced. If, on the other hand, we measure all 0s or all 1s for more than half the input state space x, we know with certainty that the function was uniform.

However, if the function were balanced, there is the possibility that it might conspire against us to fool us into thinking the function was uniform until we evaluate half plus one of the input states, requiring $O(2^n)$ oracle queries, although this will occur with exponentially low probability against the number of queries. Thus, the algorithm can be approximated with exponential asymptotic certainty in **BPP**. But considering the *worst* case rather than the *average* case, we may have to perform an exponential number of evaluations, $O(2^n)$, to know the answer with absolute certainty.

The Deutsch-Jozsa algorithm solves this rather specialised problem in the worst case using only a single quantum evaluation of the oracle.

The algorithm implementing the Deutsch-Jozsa protocol and its circuit diagram are shown in Box 28.1. The engine room of the algorithm is in the Hadamard transform, $\hat{H}^{\otimes n}$, which prepares an equal superposition of all 2^n possible input bit strings x, which are then evaluated in superposition by the oracle. To ensure unitarity, the oracle is defined to implement the transformation[1]

$$\hat{U}_f |x\rangle |y\rangle = |x\rangle |y \oplus f(x)\rangle. \tag{28.1}$$

That is, it flips bit y if $f(x) = 1$ (equivalently addition modulo 2 or an exclusive-OR (XOR) operation). An inverse Hadamard transform subsequently yields a measurement outcome with one of two possibilities:

- The 0 and 1 terms outputted from the oracle interfere perfectly constructively if the function was uniform.
- They interfere perfectly destructively if the function was balanced.

Then, with a single-shot measurement of the inverse Hadamard transformed output from the oracle we establish whether $f(x)$ was balanced or uniform with certainty. This exhibits an exponential worst-case speedup compared to a randomised classical sampling algorithm (which is classically optimal).

---

[1] Note that the seemingly more obvious choice of $\hat{U}_f |x\rangle = |f(x)\rangle$ is not unitary. This trick of introducing an additional ancillary state to enable unitary construction of arbitrary functions is a common one in quantum algorithm design.

Box 28.1 **Deutsch-Jozsa algorithm for evaluating whether the function $f(x)$ is balanced or uniform, exhibiting exponential worst-case speedup compared to the best classical BPP algorithm.**

```
function DeutschJozsa(f,n):
```

1. Prepare the $n+1$-bit state

$$|\psi\rangle_0 = |0\rangle^{\otimes n}|1\rangle. \tag{28.2}$$

2. Apply the $n+1$-bit Hadamard transform

$$|\psi\rangle_1 = \hat{H}^{\otimes(n+1)}|\psi\rangle_0$$

$$= \frac{1}{\sqrt{2^{n+1}}} \sum_{x=0}^{2^n-1} |x\rangle(|0\rangle - |1\rangle), \tag{28.3}$$

where x enumerates all n-bit binary bit strings.

3. Apply the unitary oracle, implementing the transformation

$$\hat{U}_f|x\rangle|y\rangle = |x\rangle|y \oplus f(x)\rangle, \tag{28.4}$$

where $\oplus$ denotes addition modulo 2, yielding

$$|\psi\rangle_2 = \hat{U}_f|\psi\rangle_1. \tag{28.5}$$

4. Apply another Hadamard transform,

$$|\psi\rangle_3 = \hat{H}^{\otimes n}|\psi\rangle_2. \tag{28.6}$$

5. The full evolution is thus given by

$$|\psi\rangle_{\text{out}} = (\hat{H}^{\otimes n} \otimes \hat{I}) \cdot \hat{U}_f \cdot \hat{H}^{\otimes(n+1)}|0\rangle^{\otimes n}|1\rangle. \tag{28.7}$$

6. Measure the first n qubits to determine the probability of measurement outcome $|0\rangle^{\otimes n}$.

7. This probability is given by

$$P_0 = \left| \frac{1}{2^n} \sum_{x=0}^{2^n-1} (-1)^{f(x)} \right|^2. \tag{28.8}$$

8. Depending on whether $f(x)$ was uniform or balanced, the alternating sign terms in this sum interfere constructively or destructively, yielding $P_0 = 1$ or $P_0 = 0$, respectively.

9. Thus, a single measurement outcome suffices to
determine whether $f(x)$ was balanced or uniform.

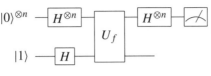

28.2 Quantum Search

The problem of finding specific entries in unstructured data spaces is a ubiquitous one. This class of *search algorithms* have amongst the broadest applicability of any class of algorithms. Computer scientists have invested excruciating man-hours[2] in organising and structuring data to minimise the resource overhead (in time and/or space) associated with extracting desired components. However, the methodology for achieving this, and the favourability of associated resource overheads, is highly dependent on the structure of the underlying data or whether there even is any. To this end, numerous data structures and algorithms have been developed, accommodating every mutation and variation of the posed problem imaginable. Often, there is a trade-off between the overheads induced in time and memory, as well as in pre-processing and data structure maintenance requirements.

For example, *hash tables* enable theoretical $O(1)$ lookup times on data with a *key–value pair* data structure. In a key–value pair each data entry (value) is tagged with a unique identifier (key) used for lookup. The value can observe any structure whatsoever, whereas the key is designed so as to minimise search times. When storing telephone numbers one might represent entries as key–value pairs, where the keys are people's names and the values their respective telephone numbers. An efficient algorithm for mapping keys to physical memory addresses would imply efficient lookup of telephone numbers by name.

In the absence of a key–value representation one might simply store data in sorted form. However, this requires pre-sorting the entire data space, which may become costly for large data sets and requires continual rearrangement whenever the data space is modified, making it computationally costly for mutable data sets.

For the end user who wishes to find data elements, the worst-case data space is one with no order or underlying structure. Suppose we want to find whether a number exists in the telephone directory but we do not know its associated name. In this instance, it can easily be seen that the best one can hope for, in terms of algorithmic runtime, is to simply look through the data space brute-force until

---

[2] Presently, most computer science research institutions are equal opportunity employers.

we find what we are looking for. It is clear that with an unstructured space of N elements this brute-force search algorithm requires on average $O(N)$ queries to find the desired entry. We call this the *unstructured search problem*.

The brute-force classical algorithm, despite already being technically 'efficient' (i.e., $O(N)$ linear runtime), could nonetheless become unwieldy for very large data sets. Google does not want to exhaustively scan their entire collection of data centres each time they want to lookup a database element. The quantum search algorithm, first presented by Grover [83], provides a solution to this problem using only $O(\sqrt{N})$ runtime (oracle queries), a quadratic enhancement. Though this falls far short of the exponential quantum enhancement one might have hoped for, which has also shown to be optimal, it is nonetheless still extremely helpful for many purposes, given the broad applications for this algorithm.

We will formulate the quantum search algorithm as an oracular algorithm, where the oracle takes as input an n-bit string and outputs 1 if the input matches the entry we are looking for, otherwise 0. This formulation of the problem makes the algorithm naturally suited to solving satisfiability problems (many of which are **NP**-complete and of great practical interest).

28.3 Quantum Simulation

The field of quantum computation was originally inspired by Feynman's observation that quantum systems cannot be efficiently classically simulated and therefore maybe computers based on quantum principles could handle this problem. Indeed they can, as was shown by [107].

It requires little imagination to recognise that the applications for quantum simulation are enormous, given the multitude of quantum systems under active investigation by researchers across countless fields.

Consider a quantum system comprised of a global Hamiltonian, which may be decomposed into smaller local interaction terms,

$$\hat{H} = \sum_{i=1}^{N} \hat{H}_i, \tag{28.9}$$

where each $\hat{H}_i$ acts on a subspace of dimension m_i within the larger system. The evolution of the entire system is given by the unitary operator

$$\hat{U} = e^{i\hat{H}t}. \tag{28.10}$$

We wish to simulate this evolution.

From the Baker-Campbell-Hausdorff lemma we can approximate this as

$$\hat{U} \approx \left(e^{i\hat{H}_1 \frac{t}{n}} \ldots e^{i\hat{H}_N \frac{t}{n}} \right)^n + O\left(\frac{t^2}{n} \right). \tag{28.11}$$

This representation effectively decomposes the global evolution into nN discretised stages of

$$\hat{U}_j = e^{i\hat{H}_j \frac{t}{n}}, \tag{28.12}$$

each of which operates on an m_i-dimensional subspace and may therefore be directly efficiently implemented as a unitary gate within the circuit model on a quantum computer.

Clearly the error terms vanish in the limit of $n \to \infty$, whereby the simulation becomes exact. However, this requires an infinite number of gates via infinitesimal discretisation. We would rather approximate the solution using a finite number of gates. It follows from Eq. (28.11) that for simulation accuracy δ, the number of discrete steps scales as

$$n = O\left(\frac{t^2}{\delta} \right), \tag{28.13}$$

which scales efficiently with the duration of time being simulated and the accuracy of the simulation.

This approach is efficient and applies to any Hamiltonian that may be decomposed into local terms as per Eq. (28.9). However, many other quantum simulation algorithms have since been described for simulating different types of quantum systems with different Hamiltonian structures [94, 34].

28.4 Integer Factorisation

By far the most influential quantum algorithm – and one of the first – is Shor's integer factorisation algorithm [165]. The problem is simply to find $x, y \in Z^+$ given $z = xy$.

Though this algorithm is known to reside in **BQP** (because it has an efficient quantum algorithm), it is strongly believed not to be **BQP**-complete. Similarly, though it is known to reside in **NP** (because it can be efficiently classically verified using simple multiplication), it is strongly believed not to be **NP**-complete, thereby placing it in the 'limbo zone' of **NP**-intermediate complexity.

This problem is of immense interest to the field of cryptography, because finding private RSA keys (Section 22.3) computationally can be reduced to this problem. An adversary with access to an efficient factoring algorithm could completely compromise RSA cryptography. For this reason, Shor's algorithm is responsible for the large investments made into quantum computing by nation states and their military and intelligence agencies!

Shor's algorithm works by first reducing integer factorisation to another problem, *period finding*. For the function

$$f(x) = a^x \bmod N, \qquad (28.14)$$

find its period $r \in Z^+$, the smallest integer such that

$$f(x) = f(x + r) \bmod N. \qquad (28.15)$$

With the ability to solve the period finding problem, an efficient classical algorithm exists for transforming this solution to a solution for factoring.

The algorithm derives its power from a quantum Fourier transform subroutine and has runtime

$$O((\log N)^2 (\log \log N)(\log \log \log N)), \qquad (28.16)$$

making it quantum efficient. By contrast, the best-known classical algorithm, the *general number field sieve*, requires time

$$O\left((e^{1.9(\log N)^{\frac{1}{3}} (\log \log N)^{\frac{2}{3}}} \right). \qquad (28.17)$$

Part VII

Cloud Quantum Computing

From the perspective of quantum computing, by far the most pressing goal for quantum networking is to facilitate *cloud quantum computing*, whereby computations can be performed over a network via a client/server model. This will be of immense importance economically, allowing very expensive quantum computers to be accessible to end users who otherwise would have been priced out of the market. This economic model is critical to the early widespread adoption of quantum computation. Networking quantum computers is also of the immense importance to capitalise off the leverage associated with unifying quantum resources as opposed to utilising them in isolation.

There are several protocols necessary to facilitate cloud quantum computing. First of all, we must have a means by which to remotely process data prepared by a host on a server(s). At the most basic level, this simply involves communicating quantum and/or classical data from a client to a single server for processing, which returns quantum or classical information to the client – *outsourced quantum computation*. In the most general case, a computation may be processed by multiple servers, each responsible for a different part of the computation – *distributed quantum computation*.

Many real-world applications for quantum computing will involve sensitive data, in terms of both the information being processed and the algorithms being employed. This necessitates encryption protocols allowing computations to be performed securely over a network, such that intercept-resend attacks are unable to infer the client's data and even the host itself is unable to do so – *homomorphic encryption* and *blind quantum computing*. These form the basic building blocks from which a secure cloud-based model for quantum computing may be constructed and economic models based on the outsourcing of computations may emerge.

The consumer of cloud quantum computing will, of course, need to be convinced that their data were processed faithfully, according to the desired algorithm. This requires *verification protocols* to allow the server to prove to the client that their data were correctly and honestly processed.

29

The Quantum Cloud

We begin by introducing the primitive building blocks for cloud quantum computing. These form the foundation for higher-level protocols to be discussed later in this part.

29.1 Outsourced Quantum Computation

Most simply, an outsourced computation involves Alice preparing either a quantum or classical input state, which she would like processed on Bob's computer. Bob performs the computation and returns either a quantum or classical state to Alice.

The algorithm, which Bob implements, could either be stipulated by Alice, in which case she is purely licensing Bob's hardware, or by Bob, in which case she is licensing his hardware and software. In the case of classical input and classical output, such an outsourced computation is trivial from a networking perspective, requiring no usage of the quantum network whatsoever. In the case of quantum input and/or output data, the quantum network will be required.

Despite the model being very simple, there may still be stringent requirements on the costs in the network. When the result of the computation is returned to Alice, there may be fidelity requirements. An approximate solution to a problem, or a computation with any logical errors whatsoever, may be useless, particularly for algorithms, which are not efficiently verifiable. For example, if Alice is attempting to factorise a large number using Shor's algorithm, a number of incorrect digits may make the the correct solution effectively impossible to determine. Or if a large satisfiability problem is being solved, almost any classical bit-flip errors will invalidate the result, requiring additional computation by Alice to resolve (which may be exponentially complex to perform).

In the case of classical communication of input and output data, we can reasonably assume error-free communication, owing to its digital nature. However, in the case of quantum communication it is inevitable that at least some degree of noise will be present. Depending on the application, this may require the client and host to jointly implement a distributed implementation of quantum error correction (QEC), whereby Alice and Bob communicate encoded states with one another, to which syndrome measurement and error correction are applied upon receipt. This will necessitate a limited amount of quantum processing to be directly available to Alice. In the case where she is completely starved of any quantum processing resources whatsoever, this may be a limiting factor. Otherwise, this type of cooperative QEC may be plausible.

29.2 Distributed Quantum Computation

The elementary model described above is very limited, because many realistic data processing applications will require multiple stages of computations to be performed, potentially by different hosts. For example, a client may need data processed using multiple proprietary algorithms owned by different hosts, and the processing will need to be distributed across the network [46].

In-Parallel and In-Series Computation

Classically, there are two main models for how a distributed computation may proceed – in *parallel* or in *series* – whereby subalgorithms are performed either side by side simultaneously or one after another in a pipeline. The two models are illustrated in Figure 29.1.

Classical parallel processing typically involves a root node, which delegates tasks to be performed in parallel by a number of child nodes and the results returned to the root node, which potentially applies an algorithm to merge the set of results before returning a final result to the client. Classical models such as Google's MAPREDUCE protocol [49] are built on this idea.

In classical computing, parallel processing is widely employed to shorten algorithmic runtimes. However, the increase in clock cycles scales only linearly with the number of nodes in the network: k-fold parallelisation yields an $\sim k$-fold speedup. For time-critical applications, such a linear improvement may already be highly beneficial, albeit costly.

The alternate scenario is in-series distributed computation, in which a computation proceeds through a pipeline of different stages, potentially performed by different hosts. This model allows a complex algorithm comprising smaller subroutines, each of which may be proprietary with different owners, to be delegated across the network. The different stages may communicate classical and/or quantum data.

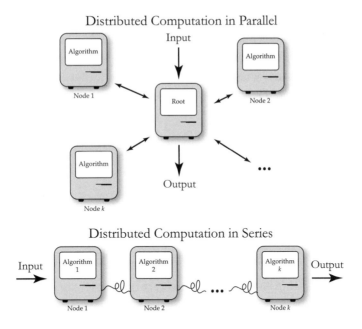

Figure 29.1 Models for distributed computation in parallel and in series. In parallel, a root node oversees the total computation, delegating tasks to child nodes, which process data independent of one another before being merged. In series the nodes sequentially process data in a pipeline of algorithmic stages.

As with the simple single-host model, if the different stages of the processing pipeline are sharing quantum data, distributed QEC will generally be necessary to protect the computation in transit. This necessarily introduces an (efficient) overhead in the number of physical qubits being communicated across the network, introducing additional bandwidth costs, which must be accommodated for in networking strategies.

Quantum Enhancement

The attractive feature of quantum computing is the potentially exponential improvement in algorithmic performance of certain tasks over their classical counterparts as the size of the computer grows. This exponential relationship implies that computation in general no longer has a simple linear trade-off as the number of participating nodes increases. In Chapter 31 we quantify this via so-called *computational scaling functions* and study its economic implications in detail.

But not every effort at distributed quantum computation will automatically exhibit the holy grail of exponential speedup. The architecture and algorithm to which it is applied must be thoughtfully designed to fully exploit the computer's quantum power. A simple adaptation of in-series or in-parallel computation may not achieve this. Rather, we must cunningly exploit quantum entanglement between nodes to perform truly distributed computation, in the sense that no instance of an algorithm is uniquely associated with any given node but is rather represented collectively across all of them.

Let us assume that we have such a carefully constructed distributed platform. Let t_c be the time required by a classical algorithm to solve a given problem and t_q the time required to solve the same problem using a quantum algorithm. In the case of algorithms exhibiting exponential quantum speedup, we will have

$$t_c = O(\exp(t_q)). \tag{29.1}$$

If we now increase the quantum processing power (i.e., number of nodes or qubits) k-fold, the equivalent classical processing time is (in the best case)

$$\begin{aligned} t_c' &= O(\exp(t_q k)) \\ &= O(\exp(t_q)^k) \\ &= O(t_c^{\,k}). \end{aligned} \tag{29.2}$$

Thus, k-fold quantum enhancement corresponds to a kth-order exponential enhancement in the equivalent classical processing time, which clearly scales much more favourably than the linear k-fold enhancement offered by classical parallelisation.

For this scaling to be possible, we expect that nodes will need to communicate via quantum rather than purely classical channels to preserve internode entanglement and mediate nonlocal gates across nodes.

Quantum MapReduce

Designing native distributed algorithms is not trivial, and architectural constraints may physically limit the allowed set of internode operations available to us. Are there any simple constructions that allow us to achieve this? We will propose an approach to parallelised quantum computation based on a direct quantum adaptation of the classical MAPREDUCE protocol.

MAPREDUCE, originally developed by Google for large-scale parallel processing, is simply an elegant formalism for parallelising classical computations. There are three stages to the protocol:

1. MAP: a root node generates k instances of an algorithm, each with different input data (or a different random seed).
2. EXECUTE: each of the k instances are executed independently on the k nodes in parallel.

3. REDUCE: all outputs are returned to the root node, collated and combined together according to some algorithm, yielding the final output of the computation.

Perhaps the simplest illustrative example is to consider the execution of a Monte Carlo simulation. Here we wish to execute a large number of instances of the same problem, each with a different random seed, and average the results to yield a statistical outcome. Here the MAP algorithm simply delegates out k copies of the same algorithm, assigning each node a different random seed, and the REDUCE algorithm needs only average their outputs. Note that the MAP and REDUCE algorithms are relatively simple, with the nodes operating in parallel doing all of the hardcore number crunching.

Taking this model, one might intuitively follow a similar approach for quantum computation, where we simply replace all of the operations with unitary processes and replace the communication links with quantum channels. Now we have a model as shown in Figure 29.2.

The goal in this construction is to make the MAP and REDUCE operations be relatively very simple – e.g., have low circuit depth – whereas the EXECUTE operations are more challenging to implement. Note that the MAP and REDUCE operations are now unitary processes, rather than being, for example, simple classical dispatch and collate operations. This means that in general the MAP operation will prepare entanglement between the EXECUTE subcomputations, and REDUCE might similarly implement nonseparable entangling measurements to measure collective properties of the joint system.

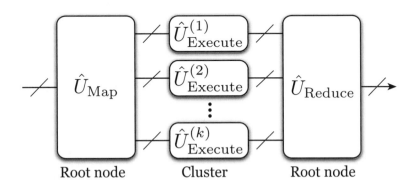

Root node   Cluster   Root node

Figure 29.2 Structure of the quantum MAPREDUCE protocol. All operations are unitary, and the MAP and REDUCE operations may be entangling in general. The EXECUTE stage is separable into a tensor product of smaller EXECUTE operations that are executed in parallel by the k nodes.

This architecture is merely a direct mapping of classical MAPREDUCE to the quantum setting. How might it be used? Consider quantum simulation, where we aim to simulate a Hamiltonian of the form

$$\hat{H}_{\text{total}} = \sum_i \hat{H}_i, \tag{29.3}$$

where each of the $\hat{H}_i$ terms are local Hamiltonians acting on orthogonal Hilbert spaces. This implies that all terms commute,

$$[\hat{H}_i, \hat{H}_j] = 0, \tag{29.4}$$

and therefore the unitaries they generate,

$$\hat{U}_j = e^{-\frac{i\hat{H}_j t}{\hbar}}, \tag{29.5}$$

have a separable tensor product structure,

$$\hat{U}_{\text{total}} = \bigotimes_i \hat{U}_i. \tag{29.6}$$

This separability lends itself directly to the tensor product structure of the EXECUTE unitaries. The MAP operation could now be a stage for preparing entangled initial states (entangled across the different subsystems), and the REDUCE operation might perform collective measurements or sampling.

Distributed Quantum Search Algorithm

The Grover quantum search algorithm can be easily parallelised by partitioning the search space and allocating a different partition to each node.

Suppose that we wish to search over the N-bit space x to find a satisfying solution to some oracle function (e.g., when solving an **NP**-complete problem),

$$x \text{ s.t. } f(x) = 1. \tag{29.7}$$

Let there be M nodes available for computation, where for simplicity we assume that M is a power of 2 (although the idea works generally for arbitrary M, albeit not as mathematically elegantly). We designate each of the nodes a $\log_2 M$-bit identification number,

$$y = [0, M - 1]. \tag{29.8}$$

We now program each node to search over a smaller search space x', which is $N - \log_2 M$ bits in length, concatenated with the node's identification number to produce the full range of x. The input to each instance of the oracle is now

$$x = x' \frown y, \tag{29.9}$$

where '$\frown$' denotes binary string concatenation.

For example, with four nodes the two-bit identification numbers are

$$y = \{00, 01, 10, 11\}. \tag{29.10}$$

If the input search space is N bits in length, then each of the nodes are assigned the search space $x' \frown y$, where x' is an $N - 2$-bit number. Within each instance, the Grover search searches over only the reduced space x', with y a constant of the instance. Figure 29.3 illustrates the circuit schematic for the simple $M = 4$ example.

It can easily be seen that this approach is compatible with the general quantum MAPREDUCE formalism, where the MAP function assigns the partitions denoted by

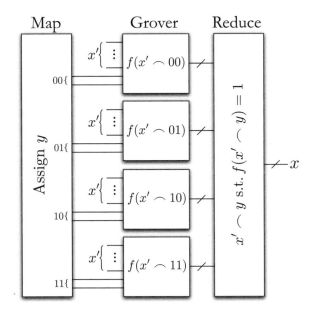

Figure 29.3 Example of a quantum MAPREDUCE protocol for implementing a distributed quantum search over four nodes. Each node performs a quantum search over the reduced space x', concatenated with the identification number of the node, which recovers the full search space x across all the nodes collectively. This effectively partitions and allocates the search-space across the nodes, which implement their reduced searches in parallel. The net speedup provided by M nodes operating in parallel scales as $O(\sqrt{M})$.

the node identification numbers y, the EXECUTE functions implement the reduced searches associated with each instance and the REDUCE function collects satisfying arguments from the instances

$$x' \text{ s.t. } f(x' \frown y) = 1 \; \forall \, y. \tag{29.11}$$

From the runtime of the Grover algorithm, it follows that the time required to solve the search problem on the initial full search space is $O(\sqrt{2^N})$, whereas the time required in the parallelised implementation is only $O(\sqrt{2^{N-\log_2 M}})$. Thus, the needed speedup is

$$O\left(\frac{\sqrt{2^N}}{\sqrt{2^{N-\log_2 M}}}\right) = O(\sqrt{M}). \tag{29.12}$$

Evidently, the net computational speedup scales as a factor of the square root of the number of nodes in the parallelised implementation. Note that this approach does not exploit entanglement between nodes and does not offer a 'quantum' (i.e., superpolynomial) speedup, because it is really just brute-force partitioning of a problem into smaller, quicker, bite-sized chunks that are attacked completely independently of one another, much like classical parallelisation.

To the contrary, unlike most quantum algorithms, whose power grows exponentially with the number of qubits (increasing returns), the distributed quantum search algorithm exhibits diminishing returns with the degree of parallelisation – the computational gain from adding one additional node to the network scales as

$$G = \sqrt{\frac{M+1}{M}}, \tag{29.13}$$

which in the large M limit asymptotes to

$$\lim_{M \to \infty} \sqrt{\frac{M+1}{M}} = 1. \tag{29.14}$$

That is, increasing the number of nodes from 1 to 2 has far greater net gain than increasing them from 100 to 101. In fact, asymptotically, the gain from adding an additional node vanishes in the limit of a high degree of parallelisation.

For this reason, parallelised implementation of a quantum search is not an example of a distributed quantum computation that achieves exponential gain with the addition of new nodes (i.e., qubits). Rather, for this specific application it is far more optimal to consolidate quantum resources into a single larger instance of a quantum search algorithm than using the quantum MAPREDUCE architecture to parallelise it.

Delocalised Computation

The cluster state, topological code and quantum random walk models for quantum computation may find themselves to be particularly well suited to distributed implementation, because they naturally reside on graphs, whose nodes need not be held locally by a single user but could instead be shared across multiple hosts with the ability for graph nodes to intercommunicate. Then only classical communication is required to complete a computation and the quantum information is not localised to any particular node.

Additionally, the entangling gates that build cluster states all commute and may be implemented simultaneously in parallel. This enables a distributed cluster state to be constructed in a 'patchwork' fashion, as shown in Figure 29.4. Now the computation is truly distributed in the sense that the computation resides collectively across the distributed cluster state, held by any number of users. No instance of an algorithm can be uniquely associated with any given node.

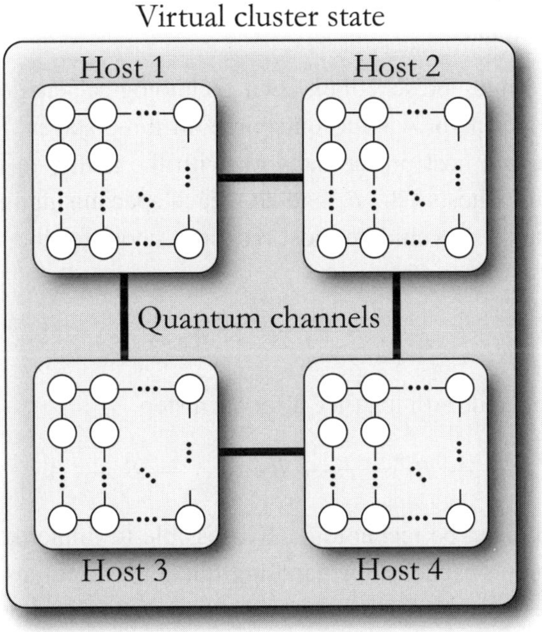

Figure 29.4 Approach for constructing distributed cluster states (or topological codes or other graph states) across multiple nodes. The quantum channels allow neighbouring clusters in the topology to be fused together, constructing a large virtual cluster state for distributed computation. The nodes could be arbitrarily separated with optically mediated interconnects to enable fusing nodes together.

This approach overlaps with the modularised approach for quantum computation discussed in Section 29.4, the difference being that in distributed cluster states the goal is to delocalise computations due to resource constraints, whereas for modularised computation the motivation is largely economical, driven by economy of scale.

29.3 Delegated Quantum Computation

Taking the notions of outsourced and distributed quantum computation to the logical extreme, we can envisage the situation where Alice has no quantum resources whatsoever (state preparation, evolution or measurement) but knows exactly what the processing pipeline should entail and who on the network has the different required quantum resources. We refer to this as *delegated quantum computation*, where the entire processing pipeline is outsourced to a series of hosts.

To illustrate this, let us consider a generic example of a quantum computation:

1. State preparation.
2. Partial measurement and feedforward.
3. Output measurement.

Each of these stages present their own technological challenges, sufficiently challenging that one might wish to outsource all three stages. However, suppose that there is no single host on the network with the ability to perform all three; rather, there are three hosts (B_1, B_2 and B_3), each specialising in just one of those tasks. In this instance, it would be most resource savvy for the network to implement the pipeline

$$A \to B_1 \to B_2 \to B_3 \to A, \qquad (29.15)$$

without going back and forth to Alice after each step,

$$A \to B_1 \to A \to B_2 \to A \to B_3 \to A. \qquad (29.16)$$

In fact, it may not even be technologically possible to implement back-and-forth to Alice if she has no capacity for handling quantum resources (i.e., the $A \leftrightarrow B$ stages are purely classical). An example of such a pipeline is shown in Figure 29.5.

This can be achieved by adding a PIPELINE field to the packet header prepared by Alice – a first in–first out queue describing the entire processing pipeline that Alice's packet (which initially contains only classical data) ought to follow through the network. Following completion of each stage of the pipeline we pop the stack and transmit the packet to the next specified host. Only at the very completion of the protocol is a packet (containing only classical data) returned to Alice.

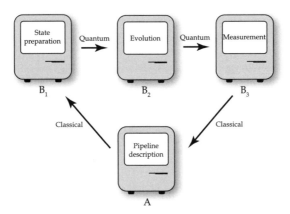

Figure 29.5 Delegated quantum computation, where each of the three computational stages (state preparation, evolution and measurement) are outsourced to the cloud without intermittent interaction with the client, A. A provides only a classical description of the processing pipeline to be implemented, each stage of which is delegated to a server specialised in that particular task. Thus, the total processing pipeline takes the form $A \rightarrow B_1 \rightarrow B_2 \rightarrow B_3 \rightarrow A$, where $A \rightarrow B_1$ and $B_3 \rightarrow A$ are classical and $B_1 \rightarrow B_2$ and $B_2 \rightarrow B_3$ are quantum channels.

Another good case study is quantum metrology using NOON states for achieving Heisenberg limited precision. Preparing NOON states is extremely challenging, and Alice may not possess the unknown phase to be measured but rather wishes a NOON state, prepared by B_1, to be provided to a third party, B_2, who applies the unknown phase and passes the resulting state to B_3, who implements the required high-efficiency parity measurements required to complete the protocol. In this case, the pipeline would take the same form as above, again with no back-and-forth communication to Alice.

Such delegated protocols will be very useful in quantum networks, where different hosts specialise in different tasks (which may be the most economically efficient model), but poor old Alice specialises in none of them, despite knowing exactly what needs to be done. This would allow an aspiring undergraduate student, who is poor (aren't they all?), to sit in his bedroom at his classical PC and implement entire distributed quantum information processing protocols in the cloud, with no quantum resources or interactions whatsoever.

29.4 Modularised Quantum Computation

How does one build a large-scale quantum computer, given the extremely daunting technological requirements and high costs? In any industry, economies of scale allow the mass production, and rapid reduction in price, of technology. To achieve

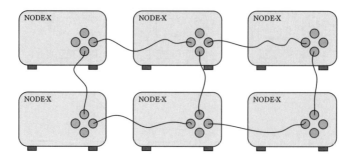

Figure 29.6 A possible physical realisation of commercially produced quantum modules, forming a 2×3 patchwork of cluster states. Each hosts a relatively small number of qubits. The nodes each have four optical interconnects, which are used to connect the modules via optical fibre. Entangling operations performed on photons shared via the interconnects create inter-module entanglement links, yielding a distributed virtual quantum computer with far more qubits. The computation is truly distributed and cooperative, in the sense that the entire computation is nonlocal, instead being collectively distributed across all the nodes, which coordinate their local operations via only classical communication. An alternate implementation is to replace the internode quantum links with Bell pair distributers. Then entanglement swapping can be employed to swap the entanglement into a link between nodes.

this, we must find a way to make quantum technologies commodity items that avoid all the hassle of customised cutting-edge labs. What we really desire is production line 'Lego for Adults', allowing ad hoc connection of *modules*, which implement small subsections of a larger computation.

We envisage that, physically, a module is a black box with optical interconnects that may be interconnected to form an arbitrary topology, yielding a physical platform as shown in Figure 29.6. The user remains oblivious to the inner workings of the modules. The modules could all be identical and just patched together differently, paving the way for their mass production and an associated quantum equivalent of Moore's Law, allowing them to become off-the-shelf commodity items over time. Then the cost of a quantum computer would simply scale linearly with its number of qubits.

The modules forming a particular computation could either be all owned by a single well-resourced operator or, alternatively, might be shared across multiple hosts, who network them remotely using entangling operations (EOs) between emitted photons.

In Section 29.2 we introduced the notion of distributed quantum computation. There the motivation was to enable a computation to be distributed across multiple servers, which either parallelise computation or process it as a pipeline in series.

An alternate direction, for economic reasons, is that it is unviable for a single server to host an entire computation. Rather, hosts will have limited capability, and performing large-scale computations will require employing a potentially large number of hosts cooperating and sharing resources with one another.[1] This can be regarded as the most general incarnation of distributed computation.

This is not the same motivation as for in-series computation, where different servers in the pipeline have different proprietary algorithms as subroutines of a larger computation. And it also differs from in-parallel computation, where multiple servers implement the same algorithm on different data, which is subsequently merged by a root node, as per, for example, a MAPREDUCE-style protocol.

Instead, the motivation is one of economics. First, individual servers will have finite resources, but there may be many of them that can be networked to cooperatively implement a larger algorithm virtually. Second, because the modules in the architecture are identical and lend themselves to mass production, one can expect more favourable economics than that offered by a provider who sells full-fledged, customised quantum computers, which do not lend themselves to the same level of mass production.

The concept of this model is best explained using the optical cluster state formalism, which lends itself naturally to this approach. A rectangular lattice graph is sufficient for universal quantum computation, even if the cluster state graph is not local (but classical communication between nodes is allowed).

Let us first assume that we wish to construct a cluster state with $n_{logical}$ logical qubits. We additionally allow each logical qubit to be the root node of a graph with a +-structure, where each branch comprises a chain of $n_{ancilla}$ ancillary physical qubits. These are sometimes referred to as *microclusters* [125]. A single microcluster collectively forms a single *module* in the topology. Our goal is to fuse modules via nearest neighbour entanglement to build up the desired distributed cluster state.

We arrange the modules to internally represent a +-topology where each node has neighbouring branches in each of the up/down/left/right directions. But we imagine the situation whereby each logical qubit, along with its respective ancillary branches, is held by a different server. Thus, the final cluster state is truly decentralised across all of the servers and, in general, entire computations cannot be performed locally.

---

[1] Even some present-day massive-scale data processing and storage protocols are implemented virtually across multiple large-scale data centres, which, for example, automatically handle geographically decentralised data redundancy and processing. Google and Amazon, for example, provide cloud services for this purpose, both employed internally and licensed out to third parties, and the Apache Cassandra project provides an open-source equivalent. The key is for the underlying protocol to abstract this away from the user, such that they interface with the data as though it were a local asset.

Using the ancillary states in the respective directions, we attempt to fuse neighbouring clusters using EOs, such as controlled-phase (CZ) gates, linear optics *fusion gates* (i.e., rotated polarising beam splitters followed by photodetection, implementing which-path erasure) [37] or atoms with a λ-configuration coupled to photons [13], which undergo which-path erasure. Importantly, using the fusion gate and which-path erasure approaches, only a single beam splitter is required to perform the EO, which only necessitates high-visibility HOM interference, mitigating the need for far more challenging interferometric (MZ) stability. This is delightful, because current leading quantum optics experiments routinely achieve HOM visibilities well in excess of 99%.

An alternate fusion strategy is not to directly communicate qubits to be bonded but instead rely off Bell pairs provided by a central authority. Each party then applies an EO between their half of the Bell pair and their target module qubit, which swaps the Bell pair entanglement onto the two respective module qubits.

When an EO is successful, we have fused two modules together, albeit potentially with some leftover ancillary states between the logical qubits. When it fails, we have lost the respective ancillary states and we attempt again using the next ancillary qubits in each of the the respective branches – a kind of REPEAT UNTIL SUCCESS strategy. The bonding only fails if all n_{ancilla} EOs fail.

Note, however, that longer ancillary arms provide more opportunity for errors to accumulate [154]. Thus, despite its tolerance against gate failure, it is nonetheless highly desirable for EOs to be as deterministic as possible to minimise the required number of ancillary qubits.

Upon successful bonding, any remaining ancillary qubits between the respective logical qubits are measured in the $\hat{Y}$ basis to remove them from the graph, whilst connecting their neighbours, leaving the two respective logical qubits as nearest neighbours in the graph. Now each module contains exactly one logical qubit, connected as desired to neighbouring modules. The relevant identities are shown in Figure 29.7. Our goal is for the entire graph to have a lattice structure, once ancillary qubits have been measured out, as illustrated in Figure 29.8.

This approach has been shown to be resource efficient [191, 125]. Let us perform a rudimentary analysis of the resource scaling of this type of approach. The probability of successfully creating an edge between two modules is

$$p_{\text{success}} = 1 - p_{\text{failure}}{}^{n_{\text{ancilla}}}, \tag{29.17}$$

where p_{success} is the probability of joining two modules, p_{failure} is the probability that a single EO fails and n_{ancilla} is the number of ancillary qubits per chain. p_{success}

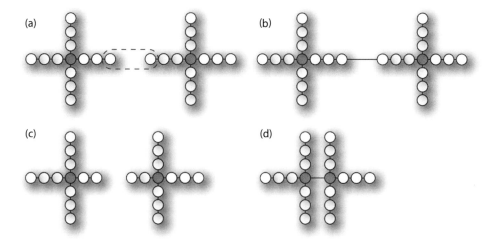

Figure 29.7 Several cluster state identities for modularised quantum computation. (a) Two cluster states with a +-topology are fused together using an EO (dashed). (b) Upon success, an edge is created between the respective qubits. (c) Upon failure, both qubits are effectively measured in the $\hat{Z}$ basis, thereby removing them, and any associated edges, from the graph. (d) Following a successful EO, the unwanted ancillary qubits may be eliminated using measurements in the $\hat{Y}$ basis, creating edges between their neighbours. If the grey qubits represent the desired logical qubits, this can be used to remove the remainder of the branches emanating from them, thereby distilling the irregular graph down to a regular lattice.

can be made arbitrarily close to unity with sufficiently long ancillary chains, the required length of which scales as

$$n_{\text{ancilla}} = \frac{\log(1 - p_{\text{success}})}{\log(p_{\text{failure}})}. \tag{29.18}$$

Now, for simplicity we will consider the preparation of linear cluster states, although these ideas can easily be extended to more complex topologies, such as 2D lattice graphs.

Let us assume that we have a 'primary' linear topology of modules, which we will incrementally attempt to 'grow' by tacking on new modules to the end. When we do so, with probability p_{success} we grow the length of the primary by 1; otherwise, we decrement it by 1. This proceeds as a random walk, with on average $2p_{\text{success}} - 1$ new qubits added to the primary per time step. Provided that this number is positive – i.e., $p_{\text{success}} > 1/2$, which can always be achieved with sufficient n_{ancilla} – the length of the primary grows linearly over time, allowing efficient state preparation.

(a)   (b)

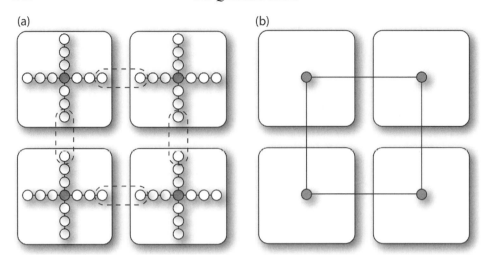

Figure 29.8 The modularised approach to scalable and economically efficient, distributed quantum computation using cluster states. The modules are all identical and can be arbitrarily patched to one another, allowing the construction of arbitrary graph topologies. Because the modules are all identical, one might hope that mass production and economy of scale will drive down the cost of modules. We consider a simple 2×2 case where each module (rounded rectangles) comprises a single logical qubit (centre of each module in grey) and a number of ancillary qubits (white in each module), which facilitate bonding the logical qubits of nearest neighbours. The preparation of the modules is performed via nearest neighbour EOs (dashed ellipses), beginning at the end of branches and working towards the root node upon each failure, until (hopefully) an EO is successful. (a) A 2×2 lattice of modules with their respective ancillary qubits. We attempt to bond the endpoints of chains using EOs. (b) Upon measuring the remaining ancillary qubits in the $\hat{Y}$ basis, only the logical qubits remain, with nearest neighbour bonds between adjacent modules, creating a distributed cluster state.

This is just a very primitive model for preparing linear cluster states, using an equally primitive INCREMENTAL strategy for constructing them using nondeterministic gates. As discussed in Section 27.2, much further work has been performed on the resource scaling of efficiently preparing cluster states of different graph topologies using different nondeterministic bonding strategies.

Of course, we have used the most simple model for modules, where each accommodates a single logical qubit. In due course, we would expect commodity modules to become far more capable and resource scaling to improve. We might envisage that each module houses a small square lattice of logical qubits, as shown in Figure 29.9, and the interconnects between them glue them together like a patchwork quilt.

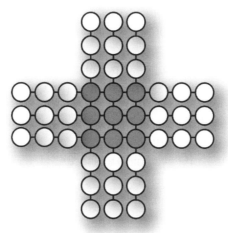

Figure 29.9 A larger cluster state module comprising a 3×3 lattice of logical qubits (grey) and dangling arms of ancillary qubits (white) in each direction for joining them to neighbouring modules. Fusing these modules enables the 'patchwork' preparation of large, distributed lattices.

29.5 Outsourced Quantum Research

Thus far we have focussed on computation as the key utility for outsourced quantum technologies, and certainly this is likely to be the dominant driving force behind quantum outsourcing. But of course not everyone wants to only solve complex algorithmic problems. Others may wish to study quantum systems themselves from the perspective of basic science research or perform precise quantum metrology.

It is foreseeable that in the context of a true quantum internet there will be a demand for not only the communication of bits and qubits but more general 'quantum assets', involving all manner of state preparation, manipulation, evolution and measurement, potentially all performed by different interconnected parties, specialising in different aspects of quantum protocols. The demand for this will extend far beyond computation.

The availability of a globe-spanning quantum satellite network brings the opportunity for fundamental quantum mechanical experiments and unprecedented length scales and velocities in the future. Satellite-to-satellite photon transfer can allow for ultra-long-distance quantum communications that are not possible on Earth due to atmospheric loss. Another unique aspect of space is that satellites move at high velocities – typically at 10^{-5} times the speed of light for low-Earth orbit (LEO) satellites. The combination of both of these effects gives a unique opportunity for performing relativistic quantum information experiments to test fundamental physics.

We anticipate that some of the first experiments will be extensions of what are already performed on Earth. For example, one can perform increasingly long space-based Bell violation tests at unprecedented distances [189]. Another possibility is to examine the speed of influence of entanglement [190]. In space, such experiments could be extended much further, giving tighter bounds. There are demanding technical hurdles that must be overcome to succeed at such experiments, such as the necessity for synchronised clocks.

In addition to examining extensions of existing experiments, the high satellite velocities can be used to perform relativistic quantum information experiments, such as entanglement tests in the presence of special and general relativity, Wheeler's delayed choice experiment and enhanced quantum metrology [95, 161, 4].

This model for quantum research could be invaluable to less-well-resourced researchers – for example, in developing nations or not-so-well-funded universities – opening up a field of experimental research previously inaccessible to them. Indeed, some private- and university-sector operators are making elementary, remotely programmable quantum information processing protocols available over the internet, bringing this type of research within reach of researchers and even curious hobbyists around the globe.

Though such early implementations fall far short of being truly reconfigurable, outsourced or delegated quantum protocols, applicable to a broad range of applications, they certainly already demonstrate the interest that such models for outsourcing are generating within the research community and the viability of further extending them.

Examples of how this type of model might be applied could include, but not be limited to, research into the following:

- Quantum information processing protocols, beyond only quantum computation, bits and qubits.
- Bose-Einstein condensates (BECs).
- Light–matter interactions.
- Quantum thermodynamics and quantum statistical mechanics.
- Quantum phase transitions.
- Quantum optics, involving all manner of quantum states of light.
- Optical interferometry.
- Providing a practical platform for university teaching and education.

In some instances, such outsourced quantum protocols might be applicable to encryption protocols, enabling highly valuable secrecy for the experiments being conducted by researchers and their hard-earned results and ideas.

29.6 The Globally Unified Quantum Cloud

In Chapter 51 we argue that in the quantum era it will be optimal to unify the world's quantum computers into a single virtual, distributed device, rather than utilising smaller individual quantum computers in isolation. This owes to the super-linear scaling in the power of a quantum computer against its number of constituent qubits, a phenomenon unique to quantum computers with no classical parallel.

This economic imperative implies that the world's many clients of quantum computing will all be interacting will a single vendor – the globally unified quantum cloud. This will create a competitive online marketplace for the licensing of timeshares in the utilisation of the unified device.

How this unified device will be managed, and by whom, is entirely open to speculation. Will a nation state or alliance of nation states monopolise it? Will a global consortium voluntarily emerge to manage the resources? Or will the whole thing be completely anarchic, potentially resulting in the fracturing of the unified device into several competing smaller ones? What policy and regulatory frameworks will emerge to oversee it?

The answers to these questions are entirely uncertain. But what is certain is that there will be an extremely high level of unification of quantum resources via the quantum internet, massively enhancing its collective computational power.

The Quantum Cloud will be far more powerful than simply licensing compute-time from a single vendor. Its collective power will be far greater than the sum of its parts.

30

Encrypted Cloud Quantum Computation

Extremely important to many high-performance data processing applications is security, because proprietary or sensitive data may be being dealt with. To address this, there are two models for encrypted, outsourced quantum computation – *homomorphic encryption* [72, 180] and *blind quantum computation* [6, 36, 15, 60, 113, 114, 114].

In both cases, Alice has secret data$^{\copyright}$, and wishes to not only ensure that an interceptor is unable to read it but that even the server performing the computation is not able to either – she trusts no one. That is, she wishes the data to be processed in encrypted form, without first requiring decryption.

The difference between the two protocols lies in the treatment of algorithms:

- Homomorphic encryption: Alice provides only the data, whereas Bob provides the processing and the algorithm it implements (which he would also like to keep to himself in general). When *any* circuit is allowed, the protocol is said to be a *fully homomorphic* encryption (FHE) protocol. Otherwise, it is a *somewhat homomorphic* encryption protocol. Although homomorphic encryption protocols have been around for a few decades in the form of privacy homomorphisms [140], classical FHE has only been described very recently [72, 180].
- Blind quantum computing: Alice provides both the algorithm *and* the data and wishes *both* to remain secret to her. It is known that universal blind *classical* computation is not possible, but universal blind *quantum* computation is.

Both of these seem like very challenging goals, yet significant developments have been made on both fronts in the quantum world, with efficient resource overheads associated with the encryption.

In the usual circuit model, blind quantum computation has been shown to be viable and optimal bounds derived. Equivalently, such protocols have been described in the cluster state model. For universal computation, such protocols necessarily require classical interaction between the client and host. However, it was shown that in some restricted (i.e., nonuniversal) models for optical quantum computation, specifically BosonSampling, quantum walks and coherent state passive linear optics, noninteractive, somewhat homomorphic encryption may be implemented.

These encryption protocols induce a resource overhead in circuit size and number of qubits involved in the computation, with efficient scaling. They deliver (at least partially) information-theoretically secure data hiding, enabling trustworthy outsourced processing of encrypted data, independent of the attack.

Part VIII

Economics and Politics

Any form of computation comes at an economic cost but also brings with it a payoff. A key consideration in any model for computation is the trade-off between the two. Because the computational power of quantum computers scales inherently differently than that of classical computers, we expect economic indicators to exhibit different scaling characteristics and dynamics also, thereby fundamentally altering the economic landscape of the post-quantum world.

We will now treat some of these economic issues in the context of a global network of unified quantum computing resources, which are then equitably time-shared. We argue in Chapter 33 that this time-shared model for quantum computation is always more computationally efficient than having distinct quantum computers operating independently in parallel, owing to the superlinear scaling in their joint computational power. Where this section provides mathematical details of various economic models, Chapter 51 provides a popular, high-level discussion surrounding these issues.

31

Classical-Equivalent Computational Power and Computational Scaling Functions

Let t be the classical-equivalent runtime of a quantum algorithm comprising n qubits – that is, how long would a given classical computer require to implement this n-qubit quantum computation? We define a *computational scaling function* characterising this relationship:

Definition 8 (Computational scaling functions) *The computational scaling function, f_{sc}, relates the number of qubits held by a quantum computer, n, and the classical-equivalent runtime, t, of the algorithm it implements,*

$$t = f_{sc}(n), \tag{31.1}$$

where f_{sc} is monotonically increasing and depends heavily on both the algorithm being implemented, as well as the architecture of the computer, including the computational model and choice of fault tolerance protocol.

The exact form of the scaling function will be specific to the algorithm being deployed[1] and the computational model (e.g., cluster states vs the circuit model, as well as choices in error correction, amongst other factors). Most notable, different quantum algorithms offer different scalings in their quantum speedup – Grover's algorithm offers only a quadratic quantum speedup, compared to the exponential speedup afforded by Shor's algorithm. Thus, the computational scaling function depends on both the hardware and software and may therefore differ between different users operating the same computer. We abstract this away and assume that all of these factors and resource overheads have been merged into the scaling function.

---

[1] For example, the *circuit depth* – i.e., number of gate applications in series – will heavily influence the number of classical steps required to simulate the circuit.

31.1 Virtual Computational Scaling Functions

If a network of quantum computers were combined into a single, larger *virtual quantum computer* using a distributed model for quantum computation, we could define a computational scaling function relationship for the virtual device:

Definition 9 (Virtual scaling function) *The joint classical-equivalent runtime of a distributed virtual quantum computation over a network is*

$$t_{\text{joint}} = f_{\text{sc}}^{\text{virtual}}(n_{\text{global}}), \tag{31.2}$$

where

$$n_{\text{global}} = \sum_{j \in \text{nodes}} n_j \tag{31.3}$$

is the total number of qubits in the network, with j summing over all nodes in the network, each of which holds n_j qubits. $f_{\text{sc}}^{\text{virtual}}$ is obtained from f_{sc} by factoring in network overheads and inefficiencies. With perfect network efficiency, $f_{\text{sc}}^{\text{virtual}} = f_{\text{sc}}$.

31.2 Combined Computational Scaling Functions

Until now we have characterised the entire network by a single scaling function. Of course, the scaling functions observed by different market participants need not all be the same, because they are functions of not only the hardware but also the participants' different algorithmic applications (i.e., software).

Consider taking a single unit of time (i.e., we are ignoring cost discounting over multiple units of time) and dividing it amongst a number of nodes, n_{nodes}, each with their own scaling function, $f_{\text{sc}}^{(i)}$. The total classical-equivalent runtime of the computation is additive, given simply by a linear combination of the classical-equivalent processing times of the individual nodes. This yields the relationship for combining scaling functions:

Definition 10 (Combined scaling functions) *The effective combined computational scaling function, $f_{\text{sc}}^{(\text{joint})}$, of a group of participants, each with their own scaling functions, $f_{\text{sc}}^{(i)}$, is given by*

$$t_{\text{joint}} = \sum_{i=1}^{n_{\text{nodes}}} \beta_i \cdot f_{\text{sc}}^{(i)}(n_{\text{global}})$$

$$= f_{\text{sc}}^{(\text{joint})}(n_{\text{global}}), \tag{31.4}$$

where β_i characterise the share of processing time allocated to each node, and for normalisation,

$$\sum_{i=1}^{n_{\text{nodes}}} \beta_i = 1. \tag{31.5}$$

Thus, the joint scaling function of the entire network is simply given by a linear combination (weighted average) of the scaling functions of the different market participants.

32

Per Qubit Computational Power

One parameter that appears ubiquitously in the upcoming economic models and warrants a definition of its own is the computational power of a quantum computer per qubit. This relates the power and size of the computer. We define this as the *per qubit computational power*:

Definition 11 (Per qubit computational power) *The per qubit computational power is defined as the computational power per qubit,*

$$\chi_{sc}(n) = \frac{f_{sc}(n)}{n}. \tag{32.1}$$

This parameter lends itself to the elegant interpretation as a cost multiplier on qubit assets, dividend and derivative prices, which warrants investigation of its scaling characteristics.

Note that in the quantum context, the computational power per qubit is not intrinsic to the qubit itself but depends on how many qubits it cooperates with, a phenomenon that does not arise in the classical context.

The key observation is that this scaling factor is constant for classical computing, where the scaling function is linear but monotonically increasing for any superlinear scaling function. For polynomial scaling functions, it has the effect of reducing the order of the polynomial by one. For exponential scaling functions, it remains exponential.

33

Time Sharing

Suppose Alice and Bob both possess expensive classical Cray supercomputers, both identical. They are both connected to the internet, so does it make sense to unify their computational resources over the network to construct a more powerful virtual machine, which they subsequently time share between themselves, or are they better off just using their own computers independently?

If there were an asymmetry in demand for computational resources, it would make perfect sense to unify computational resources, to mitigate wasting precious clock cycles. However, if they were both heavy users, always consuming every last clock cycle, it would make no difference: for a given computation, Alice and Bob could each be allocated half the processing time of the virtual supercomputer twice as powerful, or each could exploit the full processing time of their half-as-fast computers. In either case, the dollar cost of the computation is the same. This simple observation follows trivially from the linear relationship between processing power and the number of CPUs in a classical computer.

More generally, in a networked environment where time sharing of classical computational resources is applied equitably, proportionate to nodes' contribution to the network, the dollar cost per computation is (roughly, modulo parallelisation overheads) unaffected by the rest of the network. Instead, the motivation for networking computational resources is to improve efficiency by ensuring that clock cycles are not wasted but instead distributed according to demand by a scheduling algorithm, which could be market driven, for example.

However, the computational power of a quantum computer generally does not scale linearly with its number of qubits but superlinearly, often exponentially. This completely changes the economics and market dynamics of networked quantum computers. Intuitively, we expect equitable time sharing of unified quantum computational resources to offer more performance to all nodes than if they were to exclusively use their own resources in isolation. That is, the cost of a computation is reduced by resource sharing, even after time sharing.

For this reason, henceforth we will assume an environment in which owners of quantum hardware network and unify their computational power, sharing the virtual quantum computer's power between them.

34

Economic Model Assumptions

Before proceeding with explicit derivations of economic models, we state some assumptions about the dynamics of a marketplace in quantum assets. These assumptions are largely based on historical observations surrounding classical technologies that we might reasonably expect to also apply in the quantum era. However, given that the quantum marketplace is one that has not been explored in detail until now, it may be the case that some of these assumptions will require revision. Nonetheless, the general techniques we employ could readily be adapted to some relaxations and variations in these assumptions.

34.1 Efficient Markets

We make several assumptions about the efficiency of the quantum marketplace. These are largely based on the conventional efficient market hypothesis (EMH), readily taught in undergraduate ECON101 and subsequently summarily rejected upon entering ECON202. For ease of exposition, we will remain in the ECON101 classroom.

Some of these assumptions may reasonably turn out to be invalid or require revision as we learn more about upcoming quantum technologies and the trajectories their marketplace will follow. However, for ease of exposition, and the purposes of presenting some initial rudimentary, *qualitative* analyses and thought experiments, these assumptions simplify our derivations and act as a good starting point for future, more rigorous treatment (which we highly encourage!).

Given that the quantum marketplace does not actually exist yet, it is not immediately clear which assumptions are likely to be valid or not, and future, more sophisticated models will inevitably need to make more appropriate assumptions. Certainly it is no secret that in conventional settings the EMH is flawed in many respects, and some of its idealised assumptions break down in reality.

Postulate 1 (Efficient markets) *We make the following efficiency assumptions on the dynamics of the quantum marketplace:*

- *Qubits are a 'scarce' resource: there is always positive, nonzero demand for them.*
- *No wastage: quantum computational resources are always fully utilised, with no downtime.*
- *Transaction free: transaction costs are negligible, for both quantum assets and their derivatives.*
- *Negligible cost-of-carry: e.g., storage and maintenance costs are negligible.*
- *High liquidity: it is always possible to execute transactions at market rates.*
- *Perfect competition: there are no monopolies gouging prices, which are in equilibrium.*
- *Arbitrage-free: market rates for different assets and derivatives are perfectly consistent, with no opportunity for 'free money' by trading on market discrepancies.*
- *Perfect information: all market participants have complete knowledge of all market variables, including one another.*
- *Rational markets: all market participants act rationally[a] upon available information.*
- *Indefinite asset lifetime: there is no deterioration or death of quantum hardware over time.*
- *There is a risk-free rate of return (r_{rf}): the rate of growth exhibited by an investment into an optimal risk-free asset.[b]*

---

[a] *That is, with perfect economic self-interest.*

[b] *Historically these risk-free assets are taken as being US government bonds, with the bond yield being the risk-free rate of return.*

34.2 Central Mediating Authority

In Chapter 33 we argued that because of the superlinear scaling in the computational power of networked quantum computers, it will be most economically efficient to unify the world's entire collective quantum computational resources over the network and time-share their joint computational power. For this reason, we will assume that global quantum computing resources are unified, and time-shared equitably, overseen by a trusted central authority, congruent with our efficient market assumptions.

The role of the mediating authority is to perform process scheduling – equitably allocating algorithmic runtime on the virtual computer to the different network participants. This could be in the form of a state-backed authority or open market–driven alliances. In any case, the job of the authority is a relatively straightforward one, and we will assume that it induces negligible cost and computational overhead, remaining largely transparent to the end-user.

However, as discussed in Chapter 50, it may be the case that competing strategic interests will drive a wedge between the quantum resources of competitors and adversaries, partitioning them into a set of smaller networks, divided across strategic boundaries. In this instance, the arguments presented in the upcoming sections will apply to these smaller, isolated networks individually.

34.3 Network Growth

We assume that the number of qubits in the global network in the future is growing exponentially over time; i.e., the rate of progress of quantum technology will observe a Moore's law–like behaviour, as with the classical transistor.

This is a reasonable assumption based on the observation of this ubiquitous kind of behaviour in present-day technologies. Classical computing has been on a consistent exponential trajectory since the 1980s, and although it must eventually asymptote, it shows no sign of doing so in the immediate future. Quantum technologies sit at the entry point to this trajectory, and we expect it to continue for the medium term. Thus, we let the number of qubits in the network be as follows:

Postulate 2 (Network growth) *The number of qubits in the global quantum internet is growing exponentially over time as*

$$N(t) = N_0 \gamma_N{}^t, \qquad (34.1)$$

where $\gamma_N \geq 1$ characterises the rate of exponential growth in the number of qubits available to the quantum network.

The exact value of the growth rate, γ_N, is obviously unclear at such early stages in the development of the market and will ultimately be determined empirically. Although in the case of classical computing we have seen a very consistent doubling of computational power roughly every 18 months, this may very well be different for quantum technologies, owing to their fundamentally different engineering requirements (which are far more challenging in general).

34.4 Hardware Cost

Let the dollar cost of physical qubits follow Moore's law–like dynamics, decreasing exponentially with time:

Postulate 3 (Hardware cost) *The dollar cost of a single physical qubit scales inverse exponentially against time as*

$$C(t) = C_0 \gamma_C^{-t}, \qquad (34.2)$$

where $\gamma_C \geq 1$ characterises the decay rate.

This is consistent with the observed evolution of classical hardware since the beginning of the digital revolution, and it is reasonable to think that technological progress in the quantum era will follow a similar trajectory.

35

Network Power

First and foremost, with a fully interconnected quantum computational network, what is the projection of its net computational power now and into the future? This is simply obtained via the joint computational scaling function applied to projected network size:

Postulate 4 (Network power) *The combined computational power of the entire network, measured in classical-equivalent runtime is given by,*

$$P(t) = f_{\text{sc}}(n_{\text{global}})$$
$$= f_{\text{sc}}(N_0 \gamma_N{}^t). \tag{35.1}$$

36

Network Value

The simplest economic metric one might define is the collective dollar value of the entire network. That is, the product of the number of qubits on the network and the dollar cost per physical qubit at a given time:

Postulate 5 (Network value) *The dollar value of the entire network is given by*

$$V(t) = C(t)N(t)$$

$$= C_0 N_0 \left(\frac{\gamma_N}{\gamma_C} \right)^t. \qquad (36.1)$$

Note that the collective value of the network appreciates exponentially if the rate of network growth is greater than the rate of decay in the value of physical qubits; otherwise, it depreciates. At $\gamma_C = \gamma_N$ the network's dollar value remains constant over time, even if it continues expanding.

37

Rate of Return

The execution of computations typically has monetary value to the consumer. After all, they are paying hard-earned money for access to the technology!

Suppose that the owners of the quantum hardware are not running computations themselves but rather are collectively licensing out their joint compute time to end users. The hardware owners will, of course, be demanding a profit from their enterprise. The rate at which they earn back their investment into hardware via the licensing of compute time we will refer to as the rate of return (RoR), γ_{ror}. We define this as follows:

Postulate 6 (Rate of return) *The RoR is defined as*

$$e^{\gamma_{ror}(t)} = \frac{R(t)}{V(t)}, \tag{37.1}$$

where $R(t)$ is the profit made by licensing out the network's joint compute power for a single unit of time, given a present-day network value of $V(t)$.

A higher γ_{ror} implies a faster payback rate on hardware investment.[1]

---

[1] We have parametrised the RoR as an exponential for convenience when performing derivations with compounding.

38

Market Competitiveness

Recall that the risk-free rate of return (RoR) is r_{rf}. The difference between the RoR on our investment into qubit assets and the risk-free rate effectively tells us our profitability relative to a baseline zero-risk asset. This difference in turn can be interpreted as an indicator of the competitiveness or efficiency of the market – more efficient and competitive markets exhibit narrower profit windows. This yields the following figure of merit:

Postulate 7 (Market competitiveness) *The competitiveness or efficiency of the qubit market is given by the difference between the risk-free RoR and that of our physical qubits,*

$$\xi_{comp} = \gamma_{ror} - r_{rf}. \tag{38.1}$$

There are three distinct regimes for market competitiveness:

- $\xi_{comp} = 0$: *The market exhibits perfect efficiency, because price competition is so strong that profit windows have narrowed to vanishing point. There is no profit incentive to buy into or sell qubits, because they have converged with the risk-free asset.*
- $\xi_{comp} > 0$: *The market is profit-making for qubit owners. There is a profit incentive to buy ownership of physical qubits and license them out on the time-share market.*
- $\xi_{comp} < 0$: *The market is loss-making for qubit owners. Purchasing of physical qubits is disincentivised, because it is more optimal to buy into the zero-risk asset than hold qubit assets.*

The $\xi_{comp} = 0$ limit is really a fairly hypothetical regime that ought not to arise in real markets, which necessarily exhibit inefficiencies. However, highly competitive real markets will asymptote to the efficient regime, $\xi_{comp} \approx 0$.

39

Cost of Computation

As discussed in relation to combined computational scaling functions, different market participants will be executing different software applications on their share of the quantum computing resources, with differing quantum computational leverage. Because the applications differ between users, as do their computational scaling functions (f_{sc}), quantum computational leverage, so, too, does the monetary value of the computations they are performing. This yields the distinction between *subjective* and *objective* value of computation:

- Subjective value of computation: the value to an end user of a computation, measured in terms of their associated monetary profit from utilising its output, which is highly application specific.
- Objective value of computation: the cost of the physical hardware and infrastructure, which is not application specific but rather stipulated by technological and manufacturing progress.

This effectively implies that some users pay more for computation (in terms of return on investment) than others. Though the objective cost of computation is conceptually simple to model (as performed in a rudimentary fashion in Chapter 39), the subjective cost is a highly nontrivial one. It will depend heavily on the scaling function of the algorithm run by a user and, of course, the economic objectives of their computation – a quantum simulation algorithm executed by an R&D lab is likely to be of greater monetary value than an undergrad using his university's resources to execute the same task for completing an assignment!

39.1 Objective Value

In the same scenario as before, where compute time is being licensed out to end users, the hardware owner's return over a single unit of time equates to the cost of computation over that period.

Let $L(t)$ be the dollar value of utilising the network's computing resources for a single unit of time. This is obtained as the return made on the value of the network per FLOP:

Postulate 8 (Objective value of computation) *The efficient market dollar value of a computation for a single unit of time at time t per FLOP is*

$$L(t) = \frac{e^{\gamma_{\text{ror}} t} V(t)}{P(t)}$$

$$= \frac{e^{\gamma_{\text{ror}} t} C_0 \gamma_C^{-t}}{\chi_{\text{sc}}(N_0 \gamma_N^{t})}, \qquad (39.1)$$

---

which implies:

Postulate 9 (Spot price of computation) *The present-day $(t = 0)$ spot price of a computation per FLOP is*

$$L(0) = \frac{C_0}{\chi_{\text{sc}}(N_0)}. \qquad (39.2)$$

That is, the value of computations simply approximates the return on initial hardware investment, scaled by its initial computational power, as is intuitively expected.

Note that if f_{sc} scales linearly, as per classical computation, we observe a regular exponential decay in the cost of computation, consistent with the classical Moore's law. On the opposing extreme, for exponentially quantum-enhanced f_{sc}, the cost of computation decreases superexponentially with time, an economic behaviour unique to post-classical computation with no classical analogue.

The time derivative of the cost of computation is strictly negative, assuming correctness of the growth and cost postulates (Postulates 2 and 8),

$$\frac{\partial L}{\partial t} \leq 0, \qquad (39.3)$$

which implies monotonic reduction in the cost of computation over time, unless network growth and cost completely freeze ($\gamma_N = \gamma_C = 0$), in which case the cost of computation remains flat.

39.2 Subjective Value

With access to n qubits, let the subjective value extracted by user i from its execution for a unit of time be characterised by $f_{\text{sub}}^{(\text{joint})}(n)$. Then, the joint subjective value of computation (i.e., total market subjective return) follows the same form as the joint computational scaling function given in Definition 10:

Definition 12 (Subjective value of computation) *The joint subjective value of a time-share in the global network,* $f_{\text{sub}}^{(\text{joint})}$, *of a group of participants, each with their own subjective valuation functions,* $f_{\text{sub}}^{(i)}$,

$$f_{\text{sub}}^{(\text{joint})} = \sum_{i=1}^{n_{\text{nodes}}} \beta_i \cdot f_{\text{sub}}^{(i)}(n_{\text{global}}). \tag{39.4}$$

40

Arbitrage-Free Time-Sharing Model

In the context of our time-shared global network of unified quantum computers, how do we fairly and equitably allocate time shares between contributors? We now derive an elementary arbitrage-free model for equitable time sharing in such a network.

Let

$$0 \leq r_n \leq 1 \tag{40.1}$$

be the proportion of compute time allocated to a node in possession of n qubits, in a global network of n_{global} qubits. Arbitrage in the value of physical qubits will enforce the linearity constraint

$$r_{n_1+n_2} = r_{n_1} + r_{n_2}. \tag{40.2}$$

This constraint effectively mandates that 'all qubits are created equal' and two qubits are twice as valuable as one. Were, for example, a bundle of two qubits more expensive than two individual qubits purchased in isolation, a market participant could perform arbitrage and unfairly gain free compute time by buying two qubits separately, unifying them, selling the bundle, buying them back individually and repeating indefinitely until he seizes the entire network.

Additionally, we have assumed that no compute cycles are wasted – compute time is always fully utilised, as per Postulate 1. Then it follows that the time share of the combined resources of the entire network should be unity,

$$r_{n_{\text{global}}} = 1. \tag{40.3}$$

$r_{n_{\text{global}}} < 1$ would imply inefficiency via wasted clock cycles. Combining this with the linearity constraint implies the arbitrage-free time-sharing model:

Definition 13 (Arbitrage-free time-sharing model) *In an efficient market for unified quantum computing time shares, a network participant in possession of n of the entire n_{global} qubits in the network is entitled to the fraction of unified network compute time,*

$$r_n = \frac{n}{n_{\text{global}}}, \tag{40.4}$$

where

$$n_{\text{global}} = \sum_{j \in \text{nodes}} n_j, \tag{40.5}$$

is the total number of qubits in the network and

$$0 \le r_n \le 1. \tag{40.6}$$

$r_n = 1$ *iff the node has a complete monopoly over qubits; i.e., $n = n_{\text{global}}$.*

Based on this equitable model for time sharing:

Definition 14 (Time-shared computing power) *The computing power allocated to each user under the arbitrage-free time-sharing model is*

$$c_n = r_n \cdot f_{\text{sc}}(n_{\text{global}})$$
$$= n \cdot \chi_{\text{sc}}(n_{\text{global}}). \tag{40.7}$$

This model is intuitively unsurprising, because it is analogous to the case of classical computer clusters – nodes receive a time share proportional to the proportion of the hardware they are contributing to the network. However, it is important to point out that the arbitrage is taking place in the cost of physical qubits but not in terms of the dollar value of their classical-equivalent processing power, because this is in general nonlinearly related to the number of qubits. Arbitrage in computational power per se is complicated by the fact that it is a nonfungible asset that cannot be directly traded or uniquely associated with a tangible, tradable asset – its computational value is a function of other assets.

41

Problem Size Scaling Functions

The computational scaling function introduced previously expresses the power of a quantum computer in terms of its classical-equivalent runtime or, equivalently, FLOPs. However, this may not be the metric of interest when considering a computer's algorithmic power. In many situations, of far greater interest is the size of a problem instance that can be solved in a given time span. For example, the FLOPs associated with solving an instance of a 3-SAT problem grows exponentially with the number of clauses. When discussing the execution of this problem on a given computer, what we really want to know is how many clauses our device can cope with, rather than what the classical-equivalent runtime is.

This observation motivates us to reparametrise the power of quantum computers in terms of the problem size of a given algorithm to be solved. Employing the same methodology as for computational scaling functions, we define the *problem size scaling function*, which relates the size of an algorithmic problem to its classical equivalent runtime. Then equating the computational and problem size scaling function yields the follows:

Definition 15 (Problem size scaling function) *The problem size scaling function relates the size of a problem instance (s), in some arbitrary metric, to its classical-equivalent runtime (t) under a time-shared network model,*

$$t = f_{\text{size}}(s). \tag{41.1}$$

Equating this with the time-shared computational power yields

$$n \cdot \chi_{\text{sc}}(n_{\text{global}}) = f_{\text{size}}(s). \tag{41.2}$$

Isolating the problem size yields

$$s = f_{\text{size}}^{-1}(n \cdot \chi_{\text{sc}}(n_{\text{global}})). \tag{41.3}$$

We now consider several choices of scaling functions.

First let us consider the classical case of linear scaling functions (for both the computational and problem size scaling functions),

$$f_{\text{sc}}(n) = \alpha_{\text{sc}} n,$$
$$f_{\text{size}}(s) = \alpha_{\text{size}} s. \tag{41.4}$$

Solving for the problem size simply yields

$$s = \frac{n}{\alpha_{\text{size}}}$$
$$= O(1), \tag{41.5}$$

where n is regarded as a constant, and n_{global} is a variable parameter of the network. That is, the problem sizes of solvable instances are independent of the size of the external network with whom we are time sharing. This is to be expected, because these scaling functions are typical of classical computers.

For polynomial scaling functions,

$$f_{\text{sc}}(n) = n^{p_{\text{sc}}},$$
$$f_{\text{size}}(s) = s^{p_{\text{size}}}. \tag{41.6}$$

This yields problem size

$$s = (n \cdot n_{\text{global}}^{p_{\text{sc}}-1})^{\frac{1}{p_{\text{size}}}}$$
$$= O(\text{poly}(n_{\text{global}})), \tag{41.7}$$

demonstrating polynomial scaling in our solvable problem size against the size of the network.

For exponential scaling functions,

$$f_{\text{sc}}(n) = e^{\alpha_{\text{sc}} n},$$
$$f_{\text{size}}(s) = e^{\alpha_{\text{size}} s}, \tag{41.8}$$

we obtain

$$s = \log\left(n\frac{e^{\alpha_{\text{sc}} n_{\text{global}}}}{\alpha_{\text{sc}} n_{\text{global}}}\right)$$
$$= \alpha_{\text{sc}} n_{\text{global}} + \log(n) - \log(\alpha_{\text{sc}} n_{\text{global}})$$
$$= O(n_{\text{global}}), \tag{41.9}$$

demonstrating that the solvable problem size grows linearly with network size. That is, waiting for a doubling in the external network's size will also double the size of a **BQP**-complete problem that can be solved in the same time.

42

Quantum Computational Leverage

In Sections 29.2 and 29.4 we introduced distributed and modularised quantum computation. Using this as a toy model, we will now investigate the market dynamics of uniting the quantum computational resources of multiple market participants, as per an equitable time-sharing model. We envisage a model whereby network participants are contributing modules to the networked quantum computer, thereby unifying their computational power.

The ith node is contributing the fraction of the hardware r_i and receives this same proportion of compute time under the arbitrage-free time-sharing model (Definition 13). This discounts his classical-equivalent processing time to

$$\tau_i = t_{\text{joint}} \cdot r_i. \tag{42.1}$$

We are now interested in quantifying how much better off individual contributors are under this model than they were individually. Let us define the *quantum computational leverage* (QCL) of a node's quantum computer to be the ratio between their unified time-shared and individual classical-equivalent processing times,

$$\lambda_i = \frac{\tau_i}{t_i}, \tag{42.2}$$

yielding the QCL formula:

Definition 16 (Quantum computational leverage) *For the ith node, and with scaling function f_{sc}, the QCL is defined as the ratio between the unified time-shared and individual classical-equivalent algorithmic runtimes,*

$$\lambda_i = \frac{\tau_i}{t_i},$$

$$= \frac{n_i}{n_{\text{global}}} \cdot \frac{f_{sc}(n_{\text{global}})}{f_{sc}(n_i)}$$

$$= \frac{\chi_{sc}(n_{\text{global}})}{\chi_{sc}(n_i)},$$

$$\lambda_i^{\text{dB}} = 10 \log_{10}(\lambda_i), \tag{42.3}$$

where

$$n_{\text{global}} = \sum_{j \in \text{nodes}} n_j \tag{42.4}$$

is the total number of qubits in the network. The logarithmic version of the representation in decibels is simply a convenience when dealing with exponential scaling functions.

Effectively, the QCL tells us how much additional computational power we 'get for free' by consolidating with the network.

It is extremely important to note that the QCL is asymmetric, in the sense that the leverage achieved by a given node is larger than the leverage achieved by the network, upon the user joining the network (assuming that the remainder of the network comprises more qubits than the respective user).

More generally, smaller users achieve higher computational leverage from their investment into quantum hardware than larger users. Specifically,

$$\lambda_i < \lambda_j \text{ for } n_i > n_j. \tag{42.5}$$

For any superlinear scaling function we have $\lambda_i > 1 \ \forall i$, and for any linear scaling function we have $\lambda_i = 1 \ \forall i$,

$$\lambda = 1 \ \forall \ f_{sc}(n) = O(n),$$
$$\lambda > 1 \ \forall \ f_{sc}(n) > O(n). \tag{42.6}$$

For $\lambda_i > 1$ it is always computationally beneficial to all nodes to unify computational resources and time share them equitably, as per the arbitrage-free time-sharing model. Similarly, the distributed network is better off accepting them into the network, albeit to a lesser extent for a large network.

This is in contrast to classical networks, where $\lambda \approx 1$, for any number of nodes in the network (i.e., there is no leverage), and it makes no difference whether nodes unify resources or operate independently.

Finally, in the pathological case, where $\lambda_i < 1$, nodes are better off working in isolation, a situation that would only naturally arise as a result of algorithmic inefficiencies in parallelisation or distribution.

> **Definition 17 (Single-qubit QCL)** *The single-qubit QCL is the leverage associated with adding a single qubit to the network, $n = 1$, defined as*
>
> $$\lambda_{\text{qubit}} = \frac{\chi_{\text{sc}}(n_{\text{global}})}{\chi_{\text{sc}}(1)},$$
>
> $$\lambda_{\text{qubit}}^{\text{dB}} = 10 \log_{10}(\lambda_{\text{qubit}}). \tag{42.7}$$

Using our postulate for network growth (Postulate 2) yields the postulated time-dependent QCL:

> **Postulate 10 (Time-dependent QCL)** *The time-dependent QCL, based on the postulate of exponential network growth, is*
>
> $$\lambda_n(t) = \frac{\chi_{\text{sc}}(N_0 \gamma_N{}^t)}{\chi_{\text{sc}}(n)},$$
>
> $$\lambda_n^{\text{dB}}(t) = 10 \log_{10}(\lambda_n(t)). \tag{42.8}$$
>
> *The initial $(t = 0)$ time-dependent QCL reduces to the standard QCL formula.*

Note that for any superlinear scaling function, the time-dependent QCL grows exponentially over time, unlike the classical case where there is no leverage, which does not change over time (i.e., $\lambda_n(t) = 1 \ \forall n, t$).

The leverage is a function not merely of the hardware but also of the software applications running on it, each of which associated with a unique scaling function. Furthermore, it is to be reasonably anticipated that the size of the quantum internet will increase monotonically over time, yielding ever increasing leverage on the initial hardware investment by network contributors.

43

Static Computational Return

The computational leverage phenomenon clearly implies that as the global quantum network expands over time, so, too, does the computational payback on investment into network expansion or, equivalently, the cost per unit of additional classical-equivalent processing time decreases.

Because existing network participants receive leverage upon *other* participants joining the network, an investment into contributing modules has monotonically increasing computational return over time as the network expands, even if that participant ceases making further investment into the network. This is in contrast to classical networks, whereby the computational return on an investment is fixed over time.

To formalise this, consider the case where a user purchases an initial n qubits, and the global network expands over time as N_t. Then the classical-equivalent computational power of the user's fixed investment is as follows:

Definition 18 (Static computational return) *The static computational return is the classical-equivalent processing power of a user's time share proportion (n/N_t), where the user has a fixed investment of n qubits, whereas the network is allowed to expand over time arbitrarily as N_t (e.g., according to a quantum Moore's law),*

$$r_{\text{static}}(t) = \frac{n \cdot f_{\text{sc}}(N_t)}{N_t}$$

$$= n \cdot \chi_{\text{sc}}(N_t), \tag{43.1}$$

which intuitively follows as the computational power per qubit in the network times the number of qubits in our possession.

44

Forward Contract Pricing Model

Forward contracts are immensely useful in conventional markets, as a means by which to secure future use or ownership of an asset at predictable points in time. For example, farmers make heavy use of forward contracts to lock in sale of their produce before it has been harvested, such that the value is locked in in advance and the sale guaranteed, providing a very valuable hedging instrument for managing risk.

We envisage similar utility in the context of quantum computing. A company engaging in heavy use of computing power might have a need to perform certain computations at predictable points in the future. In this instance, forward contracts could be very helpful in reducing exposure to risk and guaranteeing access to the technology when needed, at a pre-agreed rate.

Now let us price forward contracts on units of computation, whereby we wish to pay today for the future use of a block of runtime on the global network.

The key observation is that a unit of computation (FLOP) does not carry over time. It must be utilised immediately and cannot be stored for future use. This simplifies the forward price of a unit of computation to simply be the future spot price, discounted by the risk-free rate of return, yielding the forward contract pricing model for quantum computing time shares:

Definition 19 (Forward contract pricing model) *The efficient market price for a forward contract in a unit of network runtime at future time T is*

$$F(T) = e^{-r_{\mathrm{rf}}T} L(T)$$

$$= \frac{e^{(\gamma_{\mathrm{ror}} - r_{\mathrm{rf}})T} C_0 \gamma_C^{-T}}{\chi_{\mathrm{sc}}(N_0 \gamma_N^{T})}. \tag{44.1}$$

Note that in the limit of $T \to 0$ this reduces to the spot price of the asset,

$$F(0) = L(0), \tag{44.2}$$

as expected.

45

Political Leverage

The asymmetry in computational leverage observed by parties of different sizes – specifically, that parties possessing a smaller number of qubits observe greater leverage than those possessing a larger number of qubits – inevitably will bring with it some power politics, with potentially interesting geo-political implications.

This asymmetry implies that in a globally unified network, were a large party to expel a small party from the network, it would be far more devastating to the computational power of the small party than the larger one. This suggests that inclusion in the global network could be a powerful tool of diplomacy in the quantum era, where threats of expulsion or resistance to inclusion is the modern-day era of gunboat diplomacy.

To quantify this, we introduce the *political leverage* quantity – the ratio between the computational leverages observed by two parties belonging to the same network. This directly quantifies the power asymmetry between them.

Definition 20 (Political leverage) *The political leverage is the ratio between the computational leverages of two parties residing on the same shared network,*

$$\gamma_{A,B} = \frac{\lambda_A}{\lambda_B}$$

$$= \frac{\chi_{sc}(n_B)}{\chi_{sc}(n_A)},$$

$$\gamma_{A,B}^{dB} = 10 \log_{10}(\gamma_{A,B}). \tag{45.1}$$

We have the trivial identity that the leverage of A against B is the inverse of the leverage of B against A,

$$\gamma_{A,B} = \gamma_{B,A}^{-1},$$

$$\gamma_{A,B}^{dB} = -\gamma_{B,A}^{dB}. \tag{45.2}$$

Note that when two parties are of equal size, there is no power asymmetry and $\gamma_{A,B} = 1$. Otherwise, when A and B are unequal, then $\gamma_{A,B} \neq 1$, indicative of power asymmetry. With linear (classical) scaling functions, the political leverage is always unity, $\gamma_{A,B} = 1$, regardless of any size asymmetry, whereas for superlinear scaling functions the political leverage diverges.

To the Machiavellian reader, this quantity can be thought of as answering the question 'If I were to expel a party from the network, how much more would it hurt them than it would hurt me?'

46

Economic Properties of the Qubit Marketplace

The development and implementation of the quantum internet will give rise to a new tradable commodity – the qubit. The pricing mechanisms associated with a qubit market were explained earlier in this part. Here we provide a broader discussion on the economic properties of such a marketplace. The are two areas of particular interest we will focus on:

1. The responsiveness of the qubit market to price fluctuations, measured by elasticity.
2. The implications of qubit market properties for broader society in terms of pricing and taxation.

46.1 The Concept of Elasticity

Elasticity as a concept is measured through percentage changes. Starting with the demand for qubits as an example, the *elasticity of demand*, E_d, is the percentage change in the quantity demanded of a good divided by the percentage change in the price of a good. Mathematically, the elasticity of demand is represented as

$$E_d = \frac{\%\Delta Q_d}{\%\Delta P}.$$ (46.1)

From this relationship, inferences about the underlying commodity can be made, summarised as follows:

- $|E_d| > 1$: the percentage change in quantity is greater than the percentage change in price and is therefore *elastic*. This indicates that demand for the asset in the market is responsive to small price changes.
- $|E_d| < 1$: there is a proportionally larger change in price, for a smaller shift in demand. This indicates that demand is less responsive to price changes and is considered *inelastic*.

- $|E_d| = 1$: we have *unit elasticity*, where the percentage change in quantity demanded is equal to the percentage change in market price.

Elasticity of demand is just one context where the concept of elasticity can be applied. Other contexts include the elasticity of supply, measuring the supply-side responsiveness to changes in price; income elasticity, which captures how the quantity of goods in the market change relative to changes in the income of consumers; and cross-price elasticity, which compares the percentage change in quantity of one good relative to the percentage change in price of another good. An example that we will come back to is how changes in the price of quantum computing may have an effect on the quantity demanded of high-performance classical computing (i.e., conventional supercomputing).

46.2 Elasticity of the Qubit Market

A number of factors will affect the elasticity of demand and supply in the qubit market. The most significant factor affecting the demand for qubits is the availability of substitutes. Given that quantum processing can efficiently solve unique problems that classical computing cannot, there are no close substitutes for qubits – a transistor is no substitute for a qubit! Consequently, elasticity of demand will be relatively inelastic: the quantity of qubits demanded will be relatively unresponsive to price fluctuations and changes, because there are no viable alternatives to substitute with.

The supply side of the equation is also initially going to be highly inelastic. However, as time progresses and technological advancements and enhancements increase the computational power of the quantum internet, the supply of qubits will become increasingly responsive to price fluctuations. However, the extent to how elastic the supply becomes over time is also going to be affected by potential for excess capacity given the exponential trajectory of the manufacture of qubits as new and improved fabrication technologies emerge – the quantum Moore's law.

47

Economic Implications

Our analysis thus far has been very theoretical. But our observations have very tangible implications in the real world. This has implications for governments, regulatory authorities, fiscal and technology policy, national security and any end users of the quantum cloud.

47.1 The Price to Pay for Isolationism

In many traditional sectors of the economy there is an economic incentive to directly compete against other market participants. However, in the quantum era the incentive is for owners of quantum computing hardware to cooperate and contribute their resources to the quantum internet rather than go it alone, as a direct consequence of superlinear leverage.

Only those hardware owners who unite with the global network will benefit from its leverage and remain competitive. Those who choose not to participate in the global network will be priced out of the market via exponentially higher cost per FLOP (assuming all other costs are equal).

This effectively taxes the cost of computation for those who fail to unify their assets with the network. And it is in the direct economic self-interest of all market participants to contribute their resources to the time-shared quantum cloud.

47.2 Taxation

Any asset, dividend, derivative or other financial instrument will inevitably be subject to taxation. Any form of taxation has multiplier effects because the cost markup is repeatedly handed from one market participant to the next, influencing the chain of supply and demand along the way. However, this multiplier and other economic consequences are highly dependent on the asset undergoing transaction – the economic implications of personal income tax are quite different to those of capital gains tax!

Computational Perspective

We now consider the effect of taxation on quantum resources, specifically in the form of a *qubit tax* – a sales tax on the purchase of physical qubits. Although this model of taxation is unlikely to be implemented as we describe, it serves as an insightful test bed for thought experiments into the qualitative implications of taxing quantum assets.

Imagine that consumers have an amount of capital available for the purchase of qubits. Let γ_T be the rate of taxation ($\gamma_T = 1$ represents no taxation, $\gamma_T > 1$ represents positive taxation and $\gamma_T < 1$ represents subsidisation). Then the cost of physical qubits is marked up by γ_T, reducing the number of qubits that can be afforded by the consumers to (assuming fixed capital available for purchasing),

$$N_{\text{tax}} = \frac{N_{\text{no tax}}}{\gamma_T}. \tag{47.1}$$

We now wish to understand how this taxation influences the computational power of the network. We define the *tax performance multiplier*:

Definition 21 (Tax performance multiplier) *The tax performance multiplier is the ratio between computational scaling functions with and without qubit taxation,*

$$M(N_{\text{tax}}) = \frac{f_{\text{sc}}(N_{\text{tax}})}{f_{\text{sc}}(N_{\text{no tax}})}$$

$$= \frac{f_{\text{sc}}(N_{\text{tax}})}{f_{\text{sc}}(N_{\text{tax}}\gamma_T)},$$

$$M^{\text{dB}}(N_{\text{tax}}) = 10\log_{10}(M(N_{\text{tax}})), \tag{47.2}$$

where the consumers have purchased N_{tax} qubits, after taxation, at a markup rate of γ_T.

The tax performance multiplier effectively gives us a factor by which computational power is depreciated under taxation. We can accomodate for other models of taxation and regulation by choosing an appropriate relationship between N_{tax}, $N_{\text{no tax}}$ and the taxation and regulatory framework.

Using our illustrative examples of computational scaling functions (linear, polynomial and exponential), the respective tax performance multipliers are given by

$$M_{\text{linear}}(N_{\text{tax}}) = \frac{1}{\gamma_T},$$

$$M_{\text{poly}}(N_{\text{tax}}) = \frac{1}{\gamma_T{}^p},$$

$$M_{\text{exp}}(N_{\text{tax}}) = e^{N_{\text{tax}}(1-\gamma_T)}. \qquad (47.3)$$

This demonstrates that the computational power of classical networks is simply inversely proportional to the rate of taxation; i.e., a linear tax performance multiplier, as we intuitively expect. And for quadratic scaling functions the dependence is inverse quadratic in the taxation rate. In both cases the multiplier is a constant factor, independent of the network size. However, for exponential scaling functions we observe an exponential dependence on both the rate of taxation and the size of the network. Note that for large networks, executing computations with exponential scaling functions, there is enormous sensitivity to variations in tax rates, yielding very high leverage in computational return by tax rates.

This implies that as the quantum network expands over time, its joint processing power decreases exponentially with the rate of taxation, yielding an ever-decreasing performance multiplier. In Chapter 51 we discuss some of the implications of this uniquely quantum phenomena.

However, taxation could also be negative, in the form of subsidisation. Evidently, even small degrees of subsidisation have a very strong impact on the performance multiplier (more pronounced than the same rate of positive taxation!). This makes subsidisation of qubit expansion highly tempting.

Policy Perspective

Irrespective of the magnitude of change, the elasticities, especially on the demand side, indicate that the qubit would be ripe for the application of a consumer-driven tax. Graphically, the imposition of a tax within a perfectly competitive market would take on the form of Figure 47.1(a).

The graph indicates a downward-sloping demand line, indicating that the lower the price, the higher the demand. The upward-sloping supply curve reflects financial incentive – the higher the price, the greater the incentive there is for firms to increase the quantity of qubits available to the market. Correlating this to elasticities, a steeper demand or supply curve indicates a higher degree of inelasticity. As such, in Figure 47.1(a), the slope of both the demand and supply curves are relatively inelastic (compared to a 45° reference line). From this, the imposition of

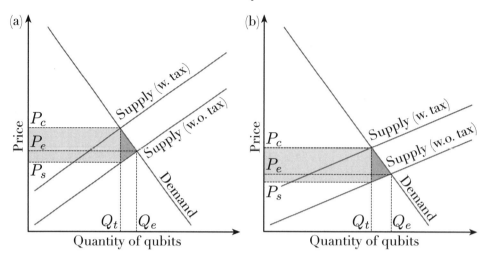

Figure 47.1 Hypothetical supply/demand curves, showing the impact of taxation on price and supply, for both inelastic (a) and elastic (b) market dynamics. Q_e and P_e are the efficient market quantity and price of the asset respectively. P_s is the price faced by suppliers, while P_c is the consumer price under taxation. Q_t is the quantity in a taxed environment. The net tax revenue collected is shaded in light teal, and the loss of market efficiency through the imposition of taxation is shaded in dark teal.

a consumer-based tax can be shown through the vertical shift of the supply curve, with the magnitude of the shift, $P_c - P_s$, indicating the per qubit value of the tax. Other important observations from the graph are the tax revenue collected (shaded in light teal), the loss of market efficiency through the imposition of a tax (shaded in dark teal) and a relatively small reduction in the quantity of qubits offered on the market from the efficient quantity, Q_e, to the quantity with the consumer tax, Q_t.

An important implication of the imposition of the tax is the share of the taxation burden. The change in price from the efficient price, P_e, to the new market price faced by consumers, P_c, is somewhat equivalent to the shift from P_e to the price point that the suppliers of qubits will receive, P_s. This means that the tax burden is likely to be equally shared between producers and consumers, which in the long term could act as a disincentive for increased production.

Figure 47.1(b), however, shows a longer-term view of the qubit market, where the supply of qubits has become more elastic in nature. That is, the quantity supplied to the market becomes more sensitive to price fluctuations. This change in elasticity results in a shift in the tax burden, with a greater proportion of the tax now being paid by consumers, with only a small shift from the efficient price, P_e, to the price suppliers will face, P_s.

From a policy implications perspective, this means that governments wanting to cash in on the new technology need to be cautious with the imposition of taxes relative to market maturity. Imposing a significant tax early on may only act as a disincentive to the development of the industry. However, once the market matures further, the imposition of a qubit tax would make strong economic sense, because there will be minimal loss of market efficiency. Importantly, such a tax on computational power alone could serve to be a relatively stable revenue generation tool for governments.

47.3 The Quantum Stock Market

In light of the distinction between subjective and objective value of computation, the question is how to reconcile this distinction in value, given the diversity of applications in the quantum marketplace. This will supersede the naïve models for cost of computation presented earlier, which were based entirely on objective value. Of course, subjective value is what people are actually willing to pay for in the real world!

This will give rise to a marketplace for tradable units of quantum computation, where the underlying asset is time shares in the global network. We refer to this as the *quantum stock market* – a marketplace subject to ordinary supply and demand, economic and, of course, psychological pressures. In a scenario where a large number of users are executing computations with high return (think the R&D lab), asset values will be traded up. Contrarily, in a scenario of low-return computations (think our poor undergrad), they will be traded down. These market forces will be highly time dependent, varying against many other factors in the economy, such as the emergence of new applications for quantum computation – the discovery of an important new algorithm could spontaneously distort the market, leading to major corrections.

The relative market value of computation will subsequently drive the direction of investment into quantum hardware, with carryover effects on future market prices. If investment stagnates, so, too, will growth in computational dividends, driving up market rates by limiting supply (assuming positive growth in demand). This will, after market adjustment, drive investment back into the system to satisfy increasing demand. Thus, despite the present uncertainty in the future dynamics of the quantum stock market, we expect this positive feedback loop to ensure consistent, ongoing investment into the quantum network and at least some marginal degree of price stability.

What is likely to arise is that most owners of quantum hardware will not be consumers but rather investors, potentially highly speculative ones, who float their resources on the quantum stock market, betting on changes in demand for computation and their associated subjective cost. This trading could involve transactions in the direct underlying asset, future contracts for locking in required computational power at future points in time or more complex derivatives. For example, an investor anticipating a surge in high-value computations is likely to invest more heavily into hardware with the expectation of an uptrend in market rates of their licensing. And their return on investment will reflect these market dynamics.

As all markets for tradable assets do, sophisticated derivative markets will inevitably emerge, whereby people can speculate on or hedge against market dynamics, taking long, short or more complex market positions, potentially in a highly leveraged manner. As discussed in Chapter 44, derivatives such as future contracts can be extremely helpful in enabling consumers to lock in future prices, creating a stable and predictable business climate. Similarly, other derivatives will enable market participants to hedge other quantum-related investments. For example, suppose an investor held a stake in an R&D lab that is highly reliant on quantum computing resources. By taking a leveraged long position on the market value for computation, he may limit losses on his R&D investment associated with the higher price (and hence lower profit) they will be paying for computation. No doubt, market manipulation and all of the usual nonsense and shenanigans will ensue.

47.4 Geographic Localisation

Because of the resource overheads associated with performing computations in a distributed manner – e.g., via the resource costs associated with long-range repeater networks – there is an economic imperative to localise quantum infrastructure to mitigate this: there is a clear economic benefit associated with housing qubits in close geographic proximity such that no long-distance quantum channels are required.

However, it is undesirable to *entirely* centralise infrastructure of *any* type, for two primary reasons:

- Geostrategic competition: competing nation states or enterprises may not want essential infrastructure to be located entirely offshore, placing them at the mercy of their strategic competitors.
- Geographical redundancy: to eliminate single points of failure, which undermine network robustness, it is desirable for infrastructure to be geographically decentralised. In present-day large-scale distributed classical platforms, geographical redundancy is a key consideration. Even though it would be most efficient if all data were completely centralised, obviating communications overheads, it would

be catastrophic if a single earthquake (or war!) could decimate the entire system. For this reason, it is desirable to distribute failure modes.

Thus, we can reasonably anticipate that the quantum internet will not evolve like the classical 'internet of things', whereby a massive number of ultra-small computational resources are scattered across the globe and networked. Rather, a relatively small number of central 'hubs' are likely to emerge, which centralise enormous computational power, interconnected via the quantum internet to form the globally unified quantum cloud.

Much like the classical internet, it is to be expected that the network that will emerge will exhibit a very hierarchical structure, following a Pareto distribution in hub size.

48

Game Theory of the Qubit Marketplace

Earlier in this part we established how a qubit market can function and how the pricing mechanisms of various derivatives may work. One of the more interesting dimensions to the development of the quantum internet is understanding *how* the cooperation between different suppliers will occur. Importantly, the expected high cost of quantum hardware means that there may be a limited number of competing vendors. For profit maximisation to occur, the most likely outcome is cooperation between them. The main question is what can be learnt from applying game theoretic techniques to the strategies available to quantum computing vendors? And, importantly, what are the implications associated with supply-side shifts, such as the imposition of taxation?

To analyse the decision-making options for qubit suppliers, game theory is an analytical tool to understand 'games' between players, where the outcome of the game is dependent on the various strategies employed by the players. The most well-known of these games is the prisoner's dilemma [133], describing the potential risks and rewards for two prisoners who are being independently questioned about a crime.

Games can be classified based on their dimension, including the number of agents, the symmetry of the utility payoff and whether they are cooperative. The prisoner's dilemma is an example of a *two-person, non-zero-sum, noncooperative game* [10]. More detailed examples of game theory have explored many of the base assumptions of this scenario, such as what if the prisoners are able to cooperate from the outset? How does this then result in maximising utility, and is cooperation always the best answer?

In the case of the quantum internet, it should be clear from the outset that there is a strong benefit associated with cooperation between vendors. Cooperative games form an important subset of the game theory domain and are the most applicable to quantum computing, where 'cooperation' translates to the unification of quantum computing resources into a larger distributed virtual quantum computer.

As indicated previously, there will be exponential enhancements in computing power associated with unification and, as such, any qubit supplier will ultimately be able to produce excess computational power through networking and cooperating with others to exploit the computational leverage phenomenon. This idea is at the centre of the analysis when applying game theory to the decision making of suppliers.

48.1 Key Concepts

For the uninitiated to economic analysis, particularly game theory, a few key concepts need to be established. This chapter by no means tries to cover these concepts in complete detail. For more detailed information we suggest referring to [173, 10, 172]. Furthermore, the analysis at this point is only descriptive in nature, as a means of establishing the space where new research can be developed. Further more rigorous investigation is encouraged. The essential concepts that we rely on taken from game theory are the following:

- *Utility*: this can be generally defined as 'the ability to assign a number (utility) to each alternative so that, for any two alternatives, one is preferred to the other if and only if the utility of the first is greater than the utility of the second' [66]. In this regard, utility is often seen as a representation of the overall benefit associated with a decision or preference. Because utility is unobservable and may be defined essentially arbitrarily, it can become subjective in nature. However, the key is not whether the values assigned to any one preference are subjective but rather whether the assigned values associated with competing decisions are comparable, so that they can be quantitatively ranked. Thus, utility is a tool for the comparison of benefits associated with decision outcomes.
- *Utility payoff matrix*: Utility values assigned to any possible decision can be formulated into a matrix that collates all of the possible decisions from a given decision-maker's perspective. In a game with two participants, this would result in a two-dimensional matrix. With n decision makers, there would be an n-dimensional matrix representation. The matrix also forms the basis for the graphical analysis undertaken throughout this section for the two-player scenarios.
- *Negotiation set*: this defines a space of possible preferences. The negotiation set was first introduced in the seminal work of [185]. At its most basic, the negotiation set forms a set of bounds for the payoff matrix. This limits possible solutions for a game to strategies that would actually see an improvement in the individual's utility above the base alternative of making no decision at all. This is also referred to as the *status quo*.

48.2 Strategies

To best develop an understanding of how a quantum internet game will be played out, we will begin by analysing the utility payoff space for classical computing. Currently we can easily define three key strategies for two market participants (X and Y), who both act as both suppliers and consumers of computational resources. The three strategies we compare are as follows:

- ISOLATION: X and Y build their own systems in isolation, which they utilise independently. There is no cooperation or networking of computational resources between them. This can be considered the status quo (S) for the players, because it represents the autarky position and defines the von Neuman and Morgenstern utility space. This is represented in Figure 48.1. Importantly, any changes to X's computational capacity has no impact on Y's. So X can take any position along the horizontal axis and there is no change in Y's utility and vice versa. The resulting *utility point* is described here as the intersecting minimum of these options, represented by the point (S_x, S_y).

- LICENSE: Either X or Y build their own systems but then licence unused compute cycles to the other. This means that the other player may still have access to their required net computational resources but essentially outsources the setup costs and ongoing infrastructure maintenance. This improves the utilisation of the system for the player who licenses out, resulting in increased efficiency, profitability and subsequently utility (under the conditions of increasing economies of scale and assuming homogenous system requirements).

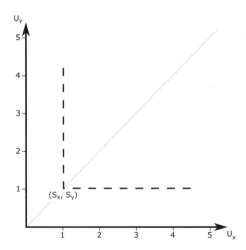

Figure 48.1 Utility space for two players with access to classical computing. The bold dashed lines show the minimum utility payoff for X and Y.

- UNIFY: The two players consolidate their computational resources into a distributed cloud computing environment, where the limitations of a single system are lifted, allowing for even better resource sharing and improvement beyond what a single system solution can provide. In this scenario, both providers will be able to collaborate such that they can meet their individual computational needs, without having to fully build independent systems as before. Importantly, though there may be a small loss in utility for one of the parties compared to the LICENSE strategy, unification allows an overall higher level of joint utility to be achieved, creating a Nash equilibrium at this point, C.

48.3 Utility Payoff Behaviour

In Table 48.1 we present an example utility payoff matrix (numbers chosen arbitrarily) between two players engaging in the above three strategies with classical computing resources.

It is important to note that these values are totally arbitrary in nature and do not have a 'real-life' interpretation, other than understanding the preferencing of the described strategies. Translating the utility payoff matrix into a graphical representation yields Figure 48.2.

Now, quantum computing uses similar strategies for possible solutions, with one key difference. Using classical computing, the relationship for the UNIFY strategy was described as additive in nature, where the computational resources of both players are accumulated additively when unified into a larger virtual computer. In the corresponding quantum UNIFY strategy, this effect is enhanced by their superlinear computational leverage. For the sake of illustration, we will now assume

Table 48.1 *Example of a payoff matrix for two classical computing vendors. '← / →' indicates the LICENSE (from/to) strategy, '+' indicates the UNIFY strategy and the off-diagonal combinations are the status quo ISOLATION strategy. (X,Y) denotes the utility to players X and Y, respectively. Note that there is some loss in net utility using LICENSE, owing to inefficiency through incurred transaction cost overheads.*

Player		X		
	Strategy	$X \to Y$	$X + Y$	$X \leftarrow Y$
Y	$X \to Y$	$(3, 1.5)$	$(1, 1)$	$(1, 1)$
	$X + Y$	$(1, 1)$	$(2.5, 2.5)$	$(1, 1)$
	$X \leftarrow Y$	$(1, 1)$	$(1, 1)$	$(1.5, 3)$

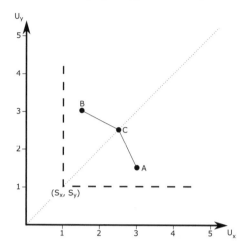

Figure 48.2 Utility payoff between two players in the classical computing environment. *A* indicates the utility combination where *X* builds the system and licenses out time to *Y*, and *B* represents the reverse. *C* is the point where both *X* and *Y* cooperate through distributed cloud computing, maximising the overall utility for both players. *A* and *B* derive identical utility values from the strategies, yielding symmetry about the diagonal axis. The lines *AC* and *BC* indicate possible solutions where mixed strategies are employed, combining components of LICENSE and UNIFY. *C* is also the Nash equilibrium, where cooperative bargaining would result in the best outcome overall.

this becomes multiplicative. Multiplicativity in computational power approximates behaviour under a UNIFY strategy when dealing with exponential scaling functions.

This is easily intuitively seen as follows. Let the computational scaling function be an arbitrary exponential,

$$f_{sc}(n) = O(\exp(n)). \tag{48.1}$$

Then it immediately follows that the scaling function obeys the identity

$$f_{sc}\left(\sum_i n_i\right) = \prod_i f_{sc}(n_i), \tag{48.2}$$

yielding the multiplicative behaviour, shown diagrammatically in the context of a distributed quantum computation in Figure 48.3. Note that in the classical case the product in Eq. (48.2) would be become a sum that is exponentially smaller in general,

$$\prod_i f_{sc}(n_i) \geq \sum_i f_{sc}(n_i). \tag{48.3}$$

Table 48.2 *Example utility payoff matrix for two players with quantum computing resources. Note the enhancement in the diagonal X+Y matrix element compared to the classical case.*

Player			X	
	Strategy	$X \to Y$	$X + Y$	$X \leftarrow Y$
	$X \to Y$	(3, 1.5)	(1, 1)	(1, 1)
Y	$X + Y$	(1, 1)	(5, 5)	(1, 1)
	$X \leftarrow Y$	(1, 1)	(1, 1)	(1.5, 3)

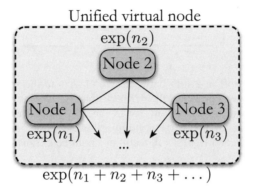

Unified virtual node

$\exp(n_2)$

Node 2

Node 1   Node 3

$\exp(n_1)$ $\exp(n_3)$

...

$\exp(n_1 + n_2 + n_3 + \dots)$

Figure 48.3 A distributed quantum computation across a number of nodes, each with n_i qubits. The computational scaling function is chosen to be exponential in form, yielding a classical-equivalent computational power of $\exp(n_i)$ for each node. However, the joint computational power of the network is given by $\exp(n_1 + n_2 + \cdots)$, which is exponentially greater than the sum of the individual computational powers, $\exp(n_1) + \exp(n_2) + \cdots$, in general.

As such, should X and Y build identical quantum computers, the hypothesised payoff matrix would become as shown in Table 48.2.

In this scenario, the ISOLATION and LICENSE strategies are assumed to yield the same utility payoffs as in the classical case. The only difference arises when X and Y UNIFY. The effect of quantum computational enhancement is to therefore amplify the cooperative elements in the utility payoff matrix, potentially by very large factors. This has the generic effect that in the quantum realm, cooperation is more highly incentivised than in the classical one. The resulting graphical representation of this payoff matrix is shown in Figure 48.4.

48.4 Cooperative Payoff Enhancement

An individual user, i, of a quantum computer operating on their own observes computational power characterised by the appropriate computational scaling function

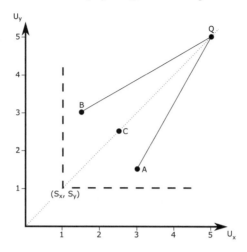

Figure 48.4 Trade-off in utility between two players in the scenario of quantum computing. Q now represents the UNIFY strategy, where there is a quantum enhancement in the utility payoff for X and Y. Note that Q offers strictly greater utility than C, the corresponding classical point. Any mixed strategy combining LICENSE and UNIFY will result in a nonlinear transition between points A/B and Q.

acting on the number qubits in their possession, $f_{sc}^{(i)}(n)$. This stipulates the utility payoff for that individual, allowing for trivial construction (and efficient mathematical representation) of multiplayer payoff matrices simply by characterising single-player payoffs independently.

Once cooperative strategies are introduced, the associated cooperative payoff matrix elements are transformed appropriately, according to the reallocation of resources and how they are collectively utilised.

We refer to this phenomenon as *cooperative payoff enhancement*, one that, depending on what cooperative techniques are employed, drastically alters the economic landscape, and its associated game-theoretic analysis and outcomes.

We will present an elementary analysis of this concept for the three different cooperation strategies introduced earlier, initially in the standard two-player context and subsequently generalised to an arbitrary multiplayer environment.

ISOLATION *Strategies*

When implementing quantum computations characterised by completely general computational scaling functions, $f_{sc}^{(X,Y)}$ (which in general can be distinct for the two players, X and Y), when using an ISOLATION strategy, the respective payoff matrix elements are simply given by

$$\text{ISOLATION} = \begin{pmatrix} f_{\text{sc}}^{(X)}(n_X) \\ f_{\text{sc}}^{(Y)}(n_Y) \end{pmatrix}, \tag{48.4}$$

where payoff matrix elements are assumed to be in units of classical-equivalent processing time[1] (i.e., FLOPs) and resource sharing is based on the methodology for arbitrage-free time-share allocation presented in Chapter 40.

LICENSE *Strategies*

Elements associated with LICENSE strategies are transformed as

$$\begin{pmatrix} f_{\text{sc}}^{(X)}(n_X) \\ f_{\text{sc}}^{(Y)}(n_Y) \end{pmatrix} \xrightarrow[\text{LICENSE}]{} \begin{pmatrix} r_{X \to X} \cdot f_{\text{sc}}^{(X)}(n_X) + r_{Y \to X} \cdot f_{\text{sc}}^{(X)}(n_Y) \\ r_{Y \to Y} \cdot f_{\text{sc}}^{(Y)}(n_Y) + r_{X \to Y} \cdot f_{\text{sc}}^{(Y)}(n_X) \end{pmatrix}, \tag{48.5}$$

where $0 \le r_{i \to j} \le 1$ denotes the proportion of i's compute time licensed to j.

This is easily logically generalised to an arbitrary multiplayer setting, in which case the transformation becomes

$$f_{\text{sc}}^{(i)}(n_i) \xrightarrow[\text{LICENSE}]{} \sum_{j=1}^{N} r_{j \to i} f_{\text{sc}}^{(i)}(n_j), \tag{48.6}$$

where there are N players, all engaging with one another using licensing only. The $r_{i \to j}$ parameters are normalised for all users such that

$$\sum_{j=1}^{N} r_{i \to j} \le 1 \ \forall i. \tag{48.7}$$

With equality, this normalisation implies perfect licensing efficiency (i.e., no overheads) and no wasted clock cycles (full utilisation). Inequality implies either inefficiency or underutilisation. Because under this strategy net computational power is conserved (at best), it might appear mindless to employ it at all, given that there is no net gain. Though this is true, there may be ulterior motives for employing it. For example, it might be employed for the purposes of load balancing across a distributed architecture or implementing arbitrage between inconsistent market pricing of computational power between nodes.

Note that when $r_{i \to j} = \delta_{i,j}$ (i.e., $r = I_N$ is the $N \times N$ identity matrix and there is no interplayer licensing) the LICENSE strategy simply reduces back to the ISOLATION strategy.

---

[1] There is nothing unique or special about using classical-equivalent computational power as our utility measure. Any other measure of 'payoff' could be equally well justified, depending on circumstance. For example, one could instead represent utility in terms of the monetary value of computational power. In that case, we simply need to transform the payoff matrix elements using the cost of computation identity presented in Postulate 8.

UNIFY *Strategies*

Elements associated with UNIFY strategies will undergo the quantum utility payoff enhancement

$$\begin{pmatrix} f_{\text{sc}}^{(X)}(n_X) \\ f_{\text{sc}}^{(Y)}(n_Y) \end{pmatrix} \xrightarrow[\text{UNIFY}]{} \begin{pmatrix} n_X \cdot \chi_{\text{sc}}^{(X)}(n_X + n_Y) \\ n_Y \cdot \chi_{\text{sc}}^{(Y)}(n_X + n_Y) \end{pmatrix} \tag{48.8}$$

from the definition for time-shared compute power given in Definition 14.

As before, we can logically generalise the payoff enhancement of a generalised UNIFY strategy to the multiplayer scenario as

$$f_{\text{sc}}^{(i)}(n_i) \xrightarrow[\text{UNIFY}]{} n_i \cdot \chi_{\text{sc}}^{(i)}(n_{\text{global}}). \tag{48.9}$$

Thus, it is evident that the enhancement in UNIFY strategies is highly dependent on the following:

• The total number of qubits held between the players,

$$n_{\text{global}} = \sum_{j=1}^{N} n_j. \tag{48.10}$$

• The proportion of the qubits held by each player,

$$r_i = \frac{n_i}{\sum_{j=1}^{N} n_j}. \tag{48.11}$$

• The respective algorithms to which the computational resources are being applied by each player, which influence the player-specific subjective scaling functions, $f_{\text{sc}}^{(i)}$, independently.

The final point is particularly noteworthy, because it implies that optimal game-theoretic outcomes are not objective but subjective and highly dependent on how players are employing their computational resources, which may be highly distinct and change dynamically over time. If one player is employing an exponential scaling function, whereas the other is only employing a polynomial one, this could completely distort the utility payoff dynamics of the game in favour of the player who would otherwise have been weaker under symmetric scaling functions. This in turn could completely alter the landscape of how users choose strategies to play optimally.

Strategic Implications

It is clear that the UNIFY strategy, in which distinct quantum computing nodes are merged via the network into a larger distributed quantum computer, works to the (potentially exponential) benefit of all contributing parties. This distorts game-theoretic analysis of network participants compared to classical computing.

On one hand, the guaranteed mutual benefit of all players directly enhances their individual compute power. In a compute-centric world, where computation equates to productivity, this directly works in the self-interest of all.

However, taking a more strategic long-term perspective, despite self-enhancement, the associated enhancement of competitors may eventuate in outcomes that work against self-interest to a sufficient extent that it outweighs this benefit. Thus, cooperative enhancement, in an appropriate strategic context, could equally be tantamount to 'adversarial enhancement' and be considered an overwhelming motivate to avoid cooperation with certain players.

Some nontechnical discussion on the implications of these observations is presented in several of the essays in Part IX.

Inefficient Markets

The utility payoff enhancement characterised by these transformations is based on the simplest of toy models, where unification is assumed to be perfectly efficient – there are no overheads (e.g., transaction or communication costs) associated with cooperation, nor are there any externalities, such as taxes or regulations. In reality, these assumptions are, of course, completely unrealistic. Thankfully, such secondary effects can be relatively easily incorporated into the model by modulating them with additional layers of transformations capturing these features.

For example, consider the unification of computational resources between two players residing in different jurisdictions, which levy import/export tariffs against one another, an externality introducing inefficiency into cooperative strategies. When expressed in terms of the monetary cost of computation, this would effectively modulate the payoffs of UNIFY matrix elements by a tariff-dependent function.

48.5 Taxation

What happens to the utility payoff matrix when taxation is imposed? The assumption that X and Y are operating in similar regulatory environments can be relaxed. What happens when X is in a more heavily taxed environment than Y or something impacts the utility achieved by the players in an asymmetric manner?

Table 48.3 *Example utility payoff matrix in the presence of taxation on quantum computers. The effect is a net deprecia-tion in achievable utility.*

Player			X	
	Strategy	$X \rightarrow Y$	$X + Y$	$X \leftarrow Y$
Y	$X \rightarrow Y$	$(2, 1.25)$	$(1, 1)$	$(1, 1)$
	$X + Y$	$(1, 1)$	$(3, 4.5)$	$(1, 1)$
	$X \leftarrow Y$	$(1, 1)$	$(1, 1)$	$(1.5, 3)$

Economically, taxation operates by transferring utility from the supplier of a good to the government. This will result in some reduction in supply. However, in the UNIFY strategy, even small reductions in supply may be exponentially mul-tiplied. The result is that all collaborative strategies will yield less utility for X, because it both reduces supply and transfers utility to government. The impact, however, is also felt by Y, because the overall joint computational power of the cloud is reduced. This results in a payoff matrix as shown in Table 48.3.

When X licenses compute time to Y, there is a reduction in utility derived by X. The small reduction in supply will also mean that there is a reduction in the available compute time available to be licensed to Y. For the LICENSE strategy, this means that there would be a shift in utility from $(3, 1.5)$ in Table 48.2 to $(2, 1.25)$ in Table 48.3. We also assume here that the imposition of the tax itself does not completely abolish any utility gains from the LICENSE strategy over the status quo.

For the UNIFY strategy, there will be a depreciation in the final utility payoff. As such, the utility payoff shows a new combination of $(3, 4.5)$, an asymmetric reduc-tion from the previous position of $(5, 5)$. Graphically, this is shown in Figure 48.5 as a shift from Q to Q'.

Two things are clear from both the matrix and its resulting graphical represen-tation. The imposition of taxation affects both the LICENSE and UNIFY strategies but unequally. For X, the imposition of taxation results in a greater loss in net utility with the UNIFY strategy. This would ultimately undermine the likelihood for X to choose this strategy. Secondly, the rate at which utility is gained through a mixed strategy (moving from A' to Q') is decreased. This implies that, with a UNIFY strategy, X would be less motivated to cooperate with Y than Y would be to cooperate with X. Q' would still lead to the maximum utility payoff but the negotiation process to achieve it will be more difficult owing to the utility asymmetry. Thus, despite both players having identical systems, the regulatory asymmetry will strongly impact the likelihood of cooperation. This has important implications for future quantum policymaking.

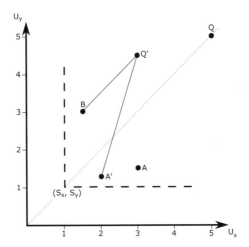

Figure 48.5 Utility payoff for two players with quantum computers employing a
UNIFY strategy under the imposition of taxation.

48.6 Resource Asymmetry

What happens with collaboration between players who have different systems? If
this assumption is relaxed, what comes of the resulting utility payoff matrix and
choice of strategies? The analysis for this assumption was presented in the previous
discussion. The effective impact of taxation is that it will limit the amount of
computational power delivered to market and capture governmental revenue from
the sale of compute time. This is analogous to what happens if X was to build a less
powerful system than Y. Through collaboration, they could still attain point Q',
shown in Figure 48.5, but this time the asymmetry arises as a design consequence
rather than a regulatory imposition. Despite X now not contributing as much to
the overall computational power of the quantum cloud, choosing the maximising
strategy of offering all available computing power to the cloud does maximise
individual utility. Furthermore, the subjective nature of the utility derivation means
that there does not have to be a change from the described payoff matrix shown
in Figure 48.4. Despite having a less powerful system, the proportional allocation
of financial rewards relative to the contribution means that both X and Y could
maintain a symmetric utility payoff outcome, like in Figure 48.4. In summary, the
two-player game shows that cooperation will result in the better outcome for both
players.

48.7 Multiplayer Games

What happens when the analysis moves beyond just two players? As established
previously, the cost and scale limitations associated with quantum computing mean

that there are likely to be limited vendors contributing to the quantum internet. However, it is also likely that there will be more than just two. Formulating the *n*-person cooperative game opens up a plethora of possibilities that go beyond the scope of this introductory discussion, but one key takeaway point from the previous analysis is that in general there is a strong motivation for computational cooperation in the quantum world. This then introduces two possible scenarios:

- There is complete cooperation between all suppliers; i.e., the global virtual quantum computer discussed in Section 29.6.
- Competing cartels develop, where for external reasons (e.g., political, geographical, ideological, strategic) there is a benefit in cooperating with a limited number of players and acting as a single supplier in direct competition with other cartels.

The first scenario is by far the most attractive, and in a world where free negotiations may take place this is clearly the option resulting in attaining both the greatest computational power for the quantum cloud and also the greatest utility maximisation for all participants involved. The end result will be similar to that shown in Figure 48.4. This implies that the quantum internet market structure would become like a natural monopoly, with diminishing long-run average costs as supply is increased. It would also mean that there is a strong case for some government involvement to prevent profit maximisation at the expense of efficiency.

However, history has shown us that another plausible outcome is the second scenario. Though there are motivations to collaborate, there are also motivations to compete. Given the almost certain involvement of government intervention in the supply of quantum processing, the formulation of regional cartels due to external factors may also be a likely outcome. In such an environment, the cartels will operate internally as multiplayer cooperative games and externally towards other conglomerates as separate multicartel competitive games [10]. This result may still result in efficient Pareto optimal outcomes at a market level but will always fall short relative to a model of complete cooperation. This scenario will naturally end with an oligopolistic market structure.

48.8 Conclusions

A game theory approach to understanding the quantum internet shows that there are strong motivations for quantum computing vendors to cooperate in order to globally maximise net utility. Furthermore, there is a strong potential to affect the possibility of cooperation through market distorting effects such as via the imposition of taxation or regulation. Finally, the two most likely market structures that will develop under the quantum internet are either a natural monopoly, where some form of regulation will be required to ensure economically efficient production, or an oligopoly, where a few cartels will compete with each other to maximise their productive output. Either way, the quantum marketplace is an extremely interesting and largely unexplored avenue for future research, a new interdisciplinary field sitting at the intersection between economics and quantum information theory. It is also one of great relevance and importance for when this technology becomes a reality.

Part IX

Essays

In this part we provide a nontechnical outlook on the future quantum internet and its implications for the benefit of the technically disinterested reader who merely wishes to grasp some of the 'big issues'. This section is in the form of a collection of short essays, requiring little or no technical background knowledge in quantum computation, quantum mechanics or mathematics.

We acknowledge that though parts of these essays are certainly highly plausible, if not certain, others are highly speculative but nonetheless based on believable although somewhat futuristic (perhaps even bordering science fiction) reasoning. We cannot predict the future, but at the very least we hope to stimulate the exchange of ideas and promote their exploration and development. After all, the great ideas of the future always begin speculatively! We encourage the reader to critically question the ideas presented in these essays and put forth their own thoughts and predictions for what the quantum future may bring and the implications it will have for humanity.

49

The Era of Quantum Supremacy

A pertinent question to ask is 'What is the timescale for useful quantum technologies? When will they be viable?' The correct answer is likely very soon.

From the perspective of classical computing, Moore's law (observation!) for the exponential growth trend in classical computing power has proven to be a very accurate one. In Figure 49.1 we illustrate the historical evolution in classical computing power and extrapolate five years into the future.

To put this into context, current-day microprocessors contain on the order of billions of single transistors. Current-day experimental quantum computers, on the other hand, contain fewer than 100 qubits. We sit at the mere very beginning of Gordon Moore's adventure through the quantum era.

Though the power of classical computers scales at most linearly with the number of transistors, the classical-equivalent power of quantum computers scales exponentially with the number of qubits (in the best-case scenario). The classical Moore's law is close to saturation – we simply cannot make transistors too much smaller than they already are![1] We therefore envisage a new quantum Moore's law, which follows a far more impressive trajectory than its classical counterpart. The point of critical mass in quantum computing will take place when the classical and quantum Moore's law extrapolations intersect, signalling the commencement of the *post-classical era* of *quantum supremacy*. Estimating this is more challenging than it sounds, because although the classical Moore's law is extremely well established with an excellent fit to an exponential trajectory, there are not yet enough data points to make a confident prediction about a quantum Moore's law, to what trajectory it best fits and at what rate it progresses, not to mention unforeseeable black swans.

---

[1] Current transistor feature sizes are on the order of several hundred atoms. Under a Moore's law prediction, we are likely to hit fundamental physical barriers in transistor size within a decade. Presumably, we can't make a transistor smaller than an atom!

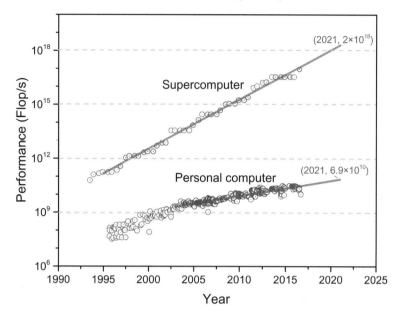

Figure 49.1 Historical trends in classical computing power for both PCs and top-end supercomputers, with an extrapolation five years into the future. The close fit to consistent exponential growth in performance over time is apparent from the logarithmic scale.

Aside from quantum computing, theoretically unbreakable quantum crypto-systems, in the form of quantum key distribution (QKD), are already techno-logically viable and are in fact commercially available off-the-shelf today, as end-to-end units connectable via fibre-optics. Recently, satellite-based QKD was demonstrated, enabling direct intercontinental QKD over thousands of kilometres. Although only a single such satellite has been demonstrated, its success implies that constellations of interconnected such satellites are inevitable in the near future, enabling point-to-point QKD between any two points on Earth. It is likely that the next space race will be the one for quantum supremacy.

As the era of post-classical quantum computation edges closer, the importance of QKD networks will intensify and along with it the demand for quantum networking infrastructure.

It is clear that humanity already sits at the precipice of harnessing quantum technologies and must act quickly to enable them to be fully exploited as they emerge in the near future.

50

The Global Virtual Quantum Computer

From the uniquely quantum phenomenon that computational power can scale exponentially with the size of a quantum computer, as opposed to the linear relationship observed in classical computing, emerges an entirely new paradigm for future supercomputing. Rather than different quantum hardware vendors competing to have the biggest and best computers, using them independently in isolation, they are incentivised to unite their resources over the network and leverage ('piggyback') off one another, forming a larger and exponentially more powerful *virtual quantum computer*, which could then be time shared between them, to the benefit of all parties. The key observation is that *all* contributing users to the network gain leverage from other users unifying their assets, irrespective of their size. In fact, this computational leverage is greater for smaller contributors than larger ones, making the benefits of this phenomenon disproportionately benefit the less-well-resourced.

Users who make an initial fixed investment into quantum computing infrastructure, which they contribute to the network but are then unable or unwilling to finance further expansion of, will nonetheless observe exponential growth in their computing power over time. That is, the computational dividend yielded by a fixed investment increases exponentially over time. This creates a very powerful economic model for investment into computational infrastructure with no classical parallel, which could be particularly valuable in developing nations or less-wealthy enterprises.

It follows that in the interests of economic efficiency, market forces will ensure that future quantum computers will *all* be networked into a single *global virtual quantum computer*, providing exponentially greater computational power to all users than what they could have afforded on their own.

Vendors of quantum compute time who do not unite with the global network will quickly be priced out of the market, owing to their reduced leverage, rendering the relative cost of their computations exponentially higher than vendors on the unified network.

This might have very interesting implications for strategic adversaries – government or private sector – competing for computational supremacy but nonetheless individually benefitting from jointly uniting their competing quantum resources. Bear in mind that using encrypted quantum computation all parties could maintain secrecy in their operations. Despite this secrecy, will the KGB and NSA really cooperate, to the benefit of both, or will the asymmetry in the computational leverage incentivise them to not unify resources and instead construct independent infrastructure?

The leverage asymmetry will be a key consideration in answering this question, because although both parties benefit on an absolute basis from unification, on a relative basis the weaker party achieves the higher computational leverage. For this reason, it is plausible that the global virtual quantum computer will fracture, dissolving into independent smaller virtual quantum computers, divided across geostrategic or competitive boundaries, with the stronger parties seceding from the union – even though they would individually benefit computationally from unification, they may not wish the weaker ones to piggyback off them, achieving greater leverage than themselves.

51

The Economics of the Quantum Internet

Quantum computers are highly likely to, at least initially, be extremely expensive and affordable outright by few. Client/server economic models based on outsourcing of computations to servers on a network will be essential to making quantum computing widely accessible. The protocols we have presented here pave the way for this type of economic model to emerge. It is paramount that the types of technologies introduced here be fully developed in time for the deployment of useful quantum computing hardware, such that they can be fully commercialised from day one of their availability, enabling widespread adoption, enhanced economy of scale and rapid proliferation.

A key question regarding the economics of the quantum internet is the extent to which it will be able to piggyback off existing optical communications infrastructure, given that networking will almost inevitably be optically mediated. We have an existing intercontinental fibre-optic backbone, as well as sophisticated satellite networks. To what extent will this existing infrastructure (or future telecom/satellite infrastructure) be able to be exploited to avoid having to rebuild the entire future quantum internet infrastructure from scratch? This is a question worth billions of dollars. We also need to factor in that given the massive driving force behind telecom technology, its cost is following a Moore's law–like trajectory of its own and what costs a billion dollars today might cost a million dollars in a decade's time. In light of this, telecom wavelength quantum optics is being hotly pursued.

Technology should benefit all of humanity, not only an elite few. In light of this, who exactly will benefit from the quantum internet? Its beauty is that it does not create a system of winners and losers. Rather, it establishes a technological infrastructure from which all can benefit, rich or poor. Well-resourced operators who can afford quantum computers, for example, will benefit from being able to license out compute time on their computers, ensuring no wasted clock cycles and maximising efficiency. The less-well-resourced will benefit in that they will have a means by which to access the extraordinary power of quantum computing on a licensed basis, facilitating access to infrastructure by those who otherwise would have been priced out of the market. This is essentially the same model as that employed by some present-day supercomputer operators, enabling small players access to supercomputing infrastructure. The quantum internet is critical to achieving the same goal in the quantum era. This could have transformative effects on the developing world in particular. And many emerging industries for whom access to quantum computation will be critical but who cannot afford them will benefit immensely from the client/server model.

Already today, even before the advent of useful post-classical quantum computers, we are seeing the emergence of the outsourced model for computation. IBM recently made an elementary 16-qubit quantum computer freely available for use via the cloud. Interested users can log in online, upload a circuit description for a quantum protocol and have it executed remotely, with the results relayed back in real time. Although still very primitive, this simple development already makes experimentation with elementary quantum protocols accessible to the poor layman, undergrad or PhD student in a developing country, people who just a few years ago would never have dreamt of being able to run their own quantum information processing experiments! This effectively opens up research opportunities to people who otherwise would have been priced out of the market entirely, unable to compete with established, well-resourced labs. Evidently, the market already recognises the importance of outsourced models for quantum computation. We encourage the impatiently curious reader to log onto the 'IBM Quantum Experience' and take a shot at designing a 16-qubit quantum protocol, without even needing to be in the same country as the quantum computer.

The quantum internet will facilitate the communication and trade of quantum assets beyond just quantum computation and cryptography. There are many uses for various hard-to-prepare quantum states; for example, in metrology, lithography or research, where outsourcing complicated state preparation would be valuable. Alternatively, performing some quantum measurements can be technologically challenging, and the ability to delegate them to someone better equipped would be desirable. The quantum internet goes beyond just quantum computing. Rather, it extends to a full range of quantum resources and protocols.

To commodify quantum computing, if constructing large-scale quantum computers were a simple matter of plug-and-play, where QuantumLego building blocks are available off-the-shelf and straightforward to assemble even for monkeys, mass production would rapidly force down prices. By arbitrarily interconnecting these boxes, large-scale quantum computers could be scaled up with demand, with a trajectory following a new quantum Moore's law, with potentially superexponential computational return.

We envisage that each of these commodity items is a black box, within which a relatively small number of qubits are held captive. Then, to build a larger quantum computer, we do not need to upgrade our boxes. Rather, we simply purchase more boxes to interconnect over the network – modularised quantum computation. This notion is tailored to graph states in particular – because a graph state can be realised by nearest neighbour interactions alone, and because all preparation stages commute with one another, they naturally lend themselves to modularised, distributed preparation.

Such an approach lends itself naturally to distributed computation, where modules may be shared across multiple users, with the economic benefit of maximising resource utilisation and the practical benefit of the end user effectively having a much larger quantum computer at their disposal.

By having a standardised architecture for optically interconnecting modules, we also somewhat 'future-proof' our hardware investment – if interfacing modules is standardised, existing hardware can be fully compatible with newer, more capable module versions. We might envision the emergence of open standards on optical interconnects and fusion protocols.

On the other hand, if quantum computers were only ever sold as specialised, room-sized, all-in-one solutions (think D-Wave), such mass production would not experience the driving force of commodified, off-the-shelf building blocks, each of which is cheap, yet frugal in its computational power alone.

Essential to existing financial markets are pricing models for physical assets. Furthermore, derivative markets increase trading liquidity, increase market efficiency, enhance price discovery and, importantly, allow risk management via hedging and the ability to lock in future prices. This is invaluable to traders of conventional commodities, and it is to be expected that it will be equally valuable to consumers of quantum resources. We have made initial steps in deriving pricing models for quantum assets and derivatives that, although they may require revision in the future real-world quantum marketplace, provide an initial qualitative understanding of quantum market dynamics.

Networked quantum computing will present new challenges for policymakers, whose fiscal policies strive to maximise economic efficiency and optimise resource allocation. Devising policies of taxation and a regulatory framework in the quantum era will require careful deliberation.

It is evident that taxation of qubits has far deeper economic implications than the taxation of other typical financial assets or classical technologies, owing to their exponential scaling characteristics. Generally speaking, taxation of an asset disincentivises its growth. But if the computational return on quantum assets grows exponentially with network size, so, too, will sensitivity to taxes that stifle it.

This will require extremely prudent consideration when designing fiscal policies in the quantum era to avoid exponential suppression of quantum-related economic activity.

Conversely, the exponential dependence on the rate of taxation could be exploited for leverage via subsidisation. It may be economically beneficial to subsidise quantum infrastructure, reaping its exponential payback, via taxation of other economic sectors, less sensitive to taxation.

The future quantum economy might be made more efficient by artificially transferring capital from low-multiplier sectors to high-multiplier quantum technologies. Or maybe the market will do this on its own accord?[1] This is a uniquely quantum consideration that never previously applied to conventional supercomputing. The onset of the quantum era may redefine our entire economic mindset and fiscal policymaking, to adapt to the unique economic idiosyncrasies of this emerging technology.

---

[1] Have faith in the invisible hand.

52

Security Implications of the Global Quantum Internet

With any new technology comes ethical considerations. Who will have access to it, and how do they plan to use it? For this reason, many developed nations have export bans or restrictions in place on 'dual-use' technologies – those that have clearly legitimate and morally justifiable uses but also nefarious ones by competitors and criminals. Nuclear technology is the obvious archetype. Quantum technologies (in particular, quantum computing and quantum cryptography) are particularly vulnerable to dual use, and for this reason are becoming subject to dual-use technology legislation, such as export controls, in some nations. In Australia, for example, legislation is being introduced criminalising the transfer of knowledge on certain quantum and cryptographic technologies to foreign nationals of certain target countries.

With a global quantum key distribution (QKD) infrastructure in place, any person on Earth with access would have uncrackable encryption at their fingertips. Though this might be welcomed by the populace of a despotic regime (or the libertarians in a democratic one), it would clearly be unwelcome for that level of secrecy and protection to be awarded to the regime itself. Similarly, criminal and terrorist organisations would be immune to government surveillance. With widespread global adoption of QKD technology, the signals intelligence agencies of nation states would become at least partially obsolete, leaving the NSA and its Five Eyes partners furious.

Quantum computing also has dual-use potential. In fact, given their ability to compromise some existing cryptographic protocols, it appears highly likely that the first useful, post-classical quantum computers will find their way into the hands of national SIGINT agencies. Of course, it does not take much imagination to see that many other quantum algorithms could be employed for sinister purposes. For this reason we are likely to see export limitations placed on quantum computer technology in the future.

Much as the internet has eliminated national electronic borders, a quantum internet employed for distributed or outsourced computation would make quantum computer technology available to hackers, criminals, terrorists and strategically competing nations. And a distributed model for computation as unregulated as the classical internet would make it near impossible to prevent.

Many falsely argue that once quantum computers become available, capable of cracking current classical cryptographic codes, the world will have transitioned to quantum-proof QKD as a replacement encryption standard and therefore that the security implications of quantum computing will not be relevant. In terms of an individual citizen's private online banking, this is largely true – who wants to read a 10-year-old bank transaction? However, when it comes to national security things are not quite so rosy. This is because major national security agencies like the NSA of the United States systematically vacuum up astronomical amounts of internet traffic and store it for future reference, knowing that one day they may be able to crack it. Thus, if the KGB had at some point in the distant past electronically communicated sensitive information that was intercepted by the NSA but unable to be cracked at the time, when sufficiently sophisticated quantum computers become available, those messages may simply be pulled from the NSA's treasure trove and trivially cracked.

Bear in mind that as recently as the late 1970s the Data Encryption Standard (DES) was a US government–approved encryption standard. However, its mere 56-bit key length is no match for a universal quantum computer. Therefore, anything stored using this encryption standard in the past had better contain information that is irrelevant by the time the quantum computers arrive, because no doubt the NSA will immediately put them to good use cracking their entire historical catalogue of stored encrypted messages.

Combined with encrypted quantum computing protocols, no one would even know what they were up to when using this awesome computing power, and what they learn, they could keep to themselves, with predictable political consequences. (see Box 52.1)

Box 52.1 A typical example of a nefarious use for cloud quantum computation.

```
function MakeAmericaGreatAgain(Putin):
```

1. A Russian bedroom hacker with no direct access to quantum technology delegates a factorisation algorithm to the cloud using homomorphic encryption.

2. The computation is physically executed on a server in the United States.
3. The result is returned to our Russian comrade.
4. The Russian uses the obtained private RSA key to hack Hillary's emails.
5. The emails are strategically leaked during the next Presidential election.
6. This swings the election in favour of Trump.
7. The NSA and FBI have no clue who was behind it, because it was homomorphically encrypted.
8. They blame Edward Snowden.
9. Fox News calls for his execution.
10. They would kick themselves if they found out the algorithm was actually executed on US soil.
11. Unless the NSA switches off the entire quantum internet, they cannot prevent it from happening again in subsequent elections.
12. return(America is Great Again).

These are all legitimate concerns. But they are very much the same ones that detractors expressed about the classical internet and strong encryption. Nonetheless, it can be said that encryption and the internet have on balance been overwhelmingly beneficial to mankind, enabling unprecedented rates of technological and economic progress. Any attempts to eliminate or undermine them could be economically catastrophic.

We take the view that the same ethical stance ought to be applied in the quantum era. Though quantum technologies clearly have dual-use potential, the magnitude of the implications they will have for scientific and technological progress overwhelms the discussed proliferation issues. No doubt, politicians will nonetheless attempt to regulate and restrict the quantum internet – that is what governments like to do. But this will inevitably fail for the same underlying reasons that it failed for the classical internet – no tech-savvy Chinese person can actually say they are hindered by the Great Firewall of China.

53

Geostrategic Quantum Politics

Computation is a commodity – perhaps the most valuable of all in the twenty-first century economy – and with any valuable, sought-after commodity comes geostrategic power play. World powers fight wars, apply sanctions and use political leverage against one another to secure access to traditional commodities essential to economic progress and competitive advantage. It is to be expected that computation will be no different.

In conventional international relations, political leverage between conflicting parties is achieved through alliances, shared common interests, threats of military action and even more sinister possibilities. How will this differ in the quantum era?

The central point to note is the computational leverage phenomena associated with the quantum internet: unification of resources is better for all. However, it is important to be cognisant that the leverage gained by parties unifying their resources with the cloud is asymmetric, biased in favour of (or against in an adversarial context) the weaker parties. That is, despite the fact that all players benefit from unification, smaller players relatively have more to gain. Though this asymmetric computational leverage may seem favourable for the weaker parties, it also places them in a compromised situation whereby the threat of a major player expelling the smaller one from the network creates asymmetric political leverage in the opposite direction. A major player will have relatively little to lose under the expulsion of a smaller player. But the smaller player could suffer immensely in the relative power of their computational assets.

This observation leads to the foreseeable possibility that future trade wars may be for computational power, with stronger parties exploiting their huge leverage over weaker parties for geopolitical objectives. Sanctions and political punishment in the quantum era may very well employ computational isolation of nation states or organisations.

It is foreseeable that the future quantum internet may become fractured along geostrategic boundaries, with players (particularly stronger ones) unwilling to provide computational leverage to strategic competitors, even though on an absolute scale they would themselves benefit, because the leverage the competitor gains may compromise their own position, for example, in cryptographic applications.

A further consideration is that the unification of quantum resources may very well require some form of central authority or marketplace to mediate the distribution and allocation of resources globally. Who will fill this role and what strategic significance it will have is hard to predict. Certainly in the case of the United Nations, the Security Council, comprising a handful of self-declared world leaders, has immense geopolitical clout, with substantial power to influence international relations across the globe. Will the United Nations, under the supervision of the Security Council or some other politicised mediating authority, oversee the international quantum marketplace, or will some self-regulating, laissez-faire, libertarian utopia emerge under the guidance of the invisible hand.

54

The Quantum Ecosystem

Associated with any new computer platform comes a hardware/software *ecosystem* that evolves around it. If we consider the release of the original iPhone and its iOS operating system, it was not just the product itself that was revolutionary but the third-party software industry that emerged surrounding it, and it was not until this software ecosystem became established on the App Store that the product realised its full potential and became truly transformative.

From the hardware perspective, it was not until interfacing standards such as USB and Wi-Fi emerged, allowing the plethora of competing hardware products to arbitrarily interconnect and interface with one another, that the hardware realised its full potential.

In the quantum era we anticipate the same phenomena to arise. What will this *quantum ecosystem* look like? Here are some of the elements that vendors might specialise in, providing compatible components for the quantum ecosystem:

- Quantum operations: vendors selling the capacity for nontrivial state preparation (e.g., Bell, NOON and GHZ states) or measurements (e.g., complex entangling syndrome measurements).
- Software subroutines and libraries: much like classical code, many quantum computations (and other quantum protocols) can be decomposed into pipelines of subroutines. There are many quantum operations that arise repeatedly (such as quantum Fourier transforms and syndrome calculations) that vendors might specialise in for outsourcing.
- Oracles: as an essential quantum software building block, oracles will become a fundamental unit for outsourcing. These oracles will store hard-corded or algorithmically generated databases or mathematical functions. For example, for use in genetic medicine, such databases could algorithmically generate tables of candidate drug compounds, or they could implement mathematical functions whose

input space is to be searched over when quantum-enhancing the solving of **NP**-complete problems.

- Interfacing: *de facto* standards will emerge for interconnecting quantum hardware units. Most notable, standards for optical interconnects will arise.
- Modularisation: arbitrarily interconnectable units will develop, allowing quantum hardware to be constructed in an ad hoc, Lego-like manner. These modules could implement small elements of a larger quantum computation, such as housing a small part of a larger graph state, communications building blocks (such as transmitters or receivers of Bell pairs) or algorithmic building blocks such as quantum Fourier transforms.
- Classical pre-, post- or intermediate-processing: quantum computation, and other quantum protocols, typically require some degree of classical pre- or post-processing. These classical operations can be highly nontrivial. For example, a novel topological quantum error correcting code might require complex encoding, decoding and feedforward operations. Determining and implementing these operations may require complicated optimisation protocols. These might be outsourced to a specialised provider.
- Classical control: many quantum protocols require intermediate classical control; i.e., feedforward. For example, in a quantum repeater network we must control the order of entangling operations and track the 'Pauli frame', a tally of the corrections accumulated by the final entangled Bell pair.
- Quantum memory: storing qubits with long decoherence lifetimes is extremely challenging using today's technology, and it is foreseeable that vendors might specialise in this particular operation, especially once error correction is built into the memory.
- Quantum error correction: any given quantum error correcting code follows a well-defined recipe for encoding, correction and decoding. Thus, it might become an example of a subroutine specialised in by a dedicated quantum error correction vendor and licensed out as a building block for embedding into larger, fault-tolerant protocols.
- Time-share licensing market: a market will emerge for the trade and allocation of time-shares on the global virtual quantum computer. Associated financial services industries will emerge around this marketplace, including secondary markets, derivative markets, managed funds in quantum infrastructure and IPO markets.

Part X

The End

55

Conclusion: The Vision of the Quantum Internet

Quantum technologies, particularly quantum computing, will truly revolutionise countless industries. With early demonstrations of key quantum technologies – such as quantum key distribution (QKD), long-distance quantum teleportation and quantum computing – becoming a reality, it is of utmost importance that networking protocols be pursued now.

We have presented an early formulation and analysis of quantum networking protocols with the vision of enabling a future quantum internet, where quantum resources can be shared and communicated in much the same way as is presently done with digital assets. Though it is hard to foresee exactly how future quantum networks will be implemented, because there are many unknowns, many of the central ideas presented here will be applicable across architectures and implementations on an ad hoc basis.

There are a number of schools of thought one might subscribe to when quantum networking. One might demand perfect data integrity and best-case network performance. But that would come at the expense of necessitating an all-powerful central authority to oversee all communications, ensuring that scheduling was absolutely perfect – a potentially very challenging optimisation problem. Or one might tolerate lost data packets or suboptimal performance at the expense of limiting applicability but with the benefit of improved flexibility and reconfigurability. Or maybe some arbitrary compromise between different costs is best. These are open questions that need not have concrete, one-size-fits-all answers. They certainly need not be answered right now.

The quantum internet will allow quantum computation to become distributed, not just outsourced. In the same way that many present-day classical algorithms are heavily parallelised and distributed across large clusters, CUDA cores or even the internet itself (e.g., the SETI project), quantum networks will allow the distribution of quantum computation across many nodes, either in parallel, in series or in a modularised fashion. This will be pivotal to achieving scalability. Keeping in mind that the classical-equivalent power of a quantum computer may grow exponentially with the number of qubits, it is highly desirable to squeeze out every last available qubit for our computations – every qubit is worth more than the last!

Combined with recent advances in homomorphic encryption and blind quantum computation, commercial models for the distribution of quantum computation will emerge, allowing computational power to be outsourced, with both client and server confident in the security of their data and proprietary algorithms. This is a notion that is challenging on classical computers but will be of utmost importance in quantum computing, where it is expected sensitive or valuable data and algorithms will often be at stake.

From the security perspective, the global quantum internet will enable an international QKD communications network with perfect secrecy, guaranteed to be information-theoretically secure by the laws of physics. This will be of immense economic and strategic benefit to commercial enterprises, governments and individuals. Classical cryptography is already a multi-billion-dollar industry worldwide. Quantum cryptography will supersede it and be of special importance in the era of quantum computers, which compromise some essential classical cryptographic protocols, such as RSA, which forms the basis of most current internet encryption, digital signatures and the Blockchain/Bitcoin protocols. Not only is quantum cryptography being pursued optically but even credit cards with embedded quantum circuitry are being actively developed to prevent fraud. Inevitably, this will require the communication between bank automats and servers to be mediated by a quantum network.

Off-the-shelf QKD systems are already available as commodity items, from vendors such as MagiQ and ID Quantique, which may be simply interconnected via an optical fibre link, thereby implementing end-to-end quantum cryptography in a modularised fashion. This is of a similar flavour to, and first technological step towards, modularised quantum computing, which would greatly enhance the economic viability and scalability of general-purpose quantum computing by paving the way for the mass production of elementary interconnectable modules as commodity items.

We have focussed our attention thus far on the application of quantum networking to quantum information processing applications, such as quantum computing and quantum cryptography. However, with plug-and-play quantum resources available over a network, one might envisage far greater applicability than just these.

Of particular interest are the implications of quantum networking to basic science research. Presently, experimental quantum physics research is limited to well-resourced labs with access to state of the art equipment. With the ability to license these assets over a network and dynamically interconnect them on an ad hoc basis, the ability to construct all manner of quantum experiments could be extended to all. An undergraduate laboratory would now have the ability to approach a host to politely borrow their state engineering technologies, send it to another with the ability to perform some evolution to that system and send it to yet another to perform measurement and analysis of the output – all from an undergrad lab equipped with nothing more than desktop PCs. This has broad implications for basic science research, opening it up to aspiring researchers across the globe, regardless of their direct access to cutting-edge tools. This will greatly expand the intellectual base for conducting quantum experimentation to the entire global scientific community, decimating the scientific monopolies controlled by a handful of world-leading, highly resourced experimental teams.

The reality is that we are only just beginning to understand the full potential for quantum technologies, and as we learn more we will inevitably find new uses for networking them. The full potential of digital electronics was never fully realised (or anticipated) until the emergence of the internet. It is to be expected that the same will hold in the quantum era, an era only in its inception.

Large-scale quantum computing may still seem a formidable and somewhat long-term challenge. But it is not likely to remain so. Once we have mastered the technological art of preparing qubits and implementing high-fidelity entangling operations between them, it is just a matter of sitting back and watching Gordon Moore perform his witchcraft, and scalability of quantum technology – and its rapid market-driven reduction in cost – will quickly ensue. The quantum internet will drive this rapid development by expanding both the supply and demand for access to this technology and through unification of computational resources allow them to massively enhance their collective computational power beyond their individual capabilities.

It is essential for the adoption and development of quantum technology that quantum networking infrastructure be sufficiently well developed that it is ready to be deployed the minute the first useful, post-classical hardware becomes available. The proliferation of the defining technology of the twenty-first century depends upon it.

References

[1] Achilles, D., Silberhorn, C., Sliwa, C., et al., 2004. Photon number resolving detection using time-multiplexing. *Journal of Modern Optics*, **51**, 1499.

[2] Aggarwal, D., Brennen, G. K., Lee, T., et al., 2017. Quantum attacks on Bitcoin, and how to protect against them. *Ledger*, **3**. https://doi.org/10.5195/ledger.2018.127

[3] Aharonov, D. and Ben-Or, M. 1997. Fault-tolerant quantum computation with constant error. *Proceedings of 29th Annual ACM Symposium on Theory of Computing*, 176–188. ACM.

[4] Ahmadi, M., Bruschi, D. E., Sabín, C., et al., 2014. Relativistic quantum metrology: Exploiting relativity to improve quantum measurement technologies. *Scientific Reports*, **4**, 4996.

[5] Aichele, T., Lvovsky, A. I. and Schiller, S. 2002. Optical mode characterization of single photons prepared by means of conditional measurements on a biphoton state. *European Physics Journal D*, **18**, 237.

[6] Arrighi, P. and Salvail, L. 2006. Blind quantum computation. *International Journal of Quantum Information*, **4**, 883.

[7] Aschauer, H., Calsamiglia, J., Hein, M., et al., 2004. Local invariants for multipartite entangled states allowing for a simple entanglement criterion. *Quantum Information & Computation*, **4**, 383.

[8] Avizienis, A. 1987. *The Evolution of Fault-Tolerant Computing*. Springer, New York.

[9] Azuma, K., Tamaki, K. and Lo, H. K. 2015. All photonic quantum repeaters. *Nature Communications*, **6**, 6787.

[10] Bacharach, M. 1976. *Economics and the Theory of Games*. Macmillan, London.

[11] Balensiefer, S., Kregor-Stickles, L. and Oskin, M. 2005. An evaluation framework and instruction set architecture for ion-trap based quantum micro-architectures. *SIGARCH Computer Architecture News*, **33**(2), 186.

[12] Banaszek, K. and Walmsley, I. 2003. Photon counting with loop detector. *Optics Letters*, **28**, 52.

[13] Barrett, S. D. and Kok, P. 2005. Efficienct high-fidelity quantum computation using matter qubits and linear optics. *Physical Review A*, **71**, 060310(R).

[14] Barrett, S. D., Rohde, P. P. and Stace, T. M. 2010. Scalable quantum computing with atomic ensembles. *New Journal of Physics*, **12**, 093032.

[15] Barz, S., Kashefi, E., Broadbent, A., et al., 2012. Demonstration of blind quantum computing. *Science*, **335**, 303.

[16] Barzanjeh, Sh., Vitali, D., Tombesi, P. and Milburn, G. J. 2011. Entangling optical and microwave cavity modes by means of a nanomechanical resonator. *Physical Review A*, **84**, 042342.

[17] Benjamin, S. C., Eisert, J. and Stace, T. M. 2005. Optical generation of matter qubit graph states. *New Journal of Physics*, **7**, 194.

[18] Bennett, C. H. 1992. Quantum cryptography using any two nonorthogonal states. *Physical Review Letters*, **68**, 3121.

[19] Bennett, C. H., Bernstein, H. J., Popescu, S., et al., 1996. Concentrating partial entanglement by local operations. *Physical Review A*, **53**, 2046.

[20] Bennett, C. H. and Brassard, G. 2014. Quantum cryptography: Public-key distribution and coin tossing. *Theoretical Computer Science*, **560** (Part 1), 7–11. doi: 10.1016/j.tcs.2014.05.025

[21] Bennett, C. H., Brassard, G., Crepeau, C., et al., 1993. Teleporting an unknown quantum state via dual classical and Einstein-Podolsky-Rosen channels. *Physical Review Letters*, **70**, 1895.

[22] Bennett, C. H., Brassard, G., Popescu, S., et al., 1996. Purification of noisy entanglement and faithful teleportation via noisy channels. *Physical Review Letters*, **76**, 722.

[23] Bennett, C. H. and DiVincenzo, D. P. 2000. Quantum information and computation. *Nature*, **404**, 247.

[24] Bennett, C. H., DiVincenzo, D. P., Smolin, J. A., et al., 1996. Mixed state entanglement and quantum error correction. *Physical Review A*, **54**, 3824.

[25] Bennett, C. H., DiVincenzo, D. P., Smolin, J. A., et al., 1996. Mixed-state entanglement and quantum error correction. *Physical Review A*, **54**, 3824.

[26] Berry, D. W. 2014. High-order quantum algorithm for solving linear differential equations. *Journal of Physics A: Mathematics & Theoretical*, **47**, 105301.

[27] Blum, S., O'Brien, C., Lauk, N., Bushev, P., et al., 2015. Interfacing microwave qubits and optical photons via spin ensembles. *Physical Review A*, **91**, 033834.

[28] Bochmann, J., Vainsencher, A., Awschalom, D. D., et al., 2013. Nanomechanical coupling between microwave and optical photons. *Nature Physics*, **9**, 712.

[29] Boruvka, O. 1926. About a certain minimal problem. *O Prace mor. prirodoved. spol. v Brne III*, **3**, 37.

[30] Branning, D., Grice, W., Erdmann, R., et al., 2000. Interferometric technique for engineering indistinguishability and entanglement of photon pairs. *Physical Review A*, **62**, 013814.

[31] Brattke, S., Varcoe, B. T. H. and Walther, H. 2001. Generation of photon number states on demand via cavity quantum electrodynamics. *Physical Review Letters*, **86**, 3534.

[32] Bratzik, S., Abruzzo, S., Kampermann, H., et al., 2013. Quantum repeaters and quantum key distribution: The impact of entanglement distillation on the secret-key rate. *Physical Review A*, **86**, 062335.

[33] Braunstein, S. L. and Mann, A. 1995. Measurement of the Bell operator and quantum teleportation. *Physical Review A*, **51**, R1727.

[34] Brennen, G. K., Rohde, P., Sanders, B. C., et al., 2015. Multi-scale quantum simulation of quantum field theory using wavelets. *Physical Review A*, **92**, 032315.

[35] Briegel, H. J., Dür, W., Cirac, J. I., et al., 1998. Quantum repeaters: The role of imperfect local operations in quantum communication. *Physical Review Letters*, **81**, 5932.

[36] Broadbent, A., Fitzsimons, J. and Kashefi, E. 2009. Universal blind quantum computation. Page 517 of: *IEEE Symposium on Foundations of Computer Science (FOCS)*, Vol. 50. IEEE. doi: 10.1109/FOCS.2009.36

[37] Browne, D. E. and Rudolph, T. 2005. Resource-efficient linear optics quantum computation. *Physical Review Letters*, **95**, 010501.

[38] Brunel, C., Lounis, B., Tamarat, P., et al., 1999. Triggered source of single photons based on controlled single molecule fluorescence. *Physical Review Letters*, **83**, 2722.

[39] Cahill, K. E. and Glauber, R. J. 1969. Density operators and quasiprobability distributions. *Physical Review*, **177**, 177.

[40] Calderbank, A. R. and Shor, P. W. 1996. Good quantum error-correcting codes exist. *Physical Review A*, **54**, 1098.

[41] Campbell, E. T., Fitzsimons, J., Benjamins, S. C., et al., 2007. Adaptive strategies for graph state growth in the presence of monitored errors. *Physical Review A*, **75**, 042303.

[42] Campbell, E. T., Fitzsimons, J., Benjamin, S. C., et al., 2007. Efficient growth of complex graph states via imperfect path erasure. *New Journal of Physics*, **9**, 196.

[43] Childress, L., Taylor, J. M., Sørensen, A. S., et al., 2006. Fault-tolerant quantum communication based on solid-state photon emitters. *Physical Review Letters*, **96**, 070504.

[44] Chou, C. W., de Riedmatten, H., Felinto, D., et al., 2005. Measurement-induced entanglement for excitation stored in remote atomic ensembles. *Nature*, **438**, 828.

[45] Chuang, I. L. and Nielsen, M. A. 1997. Prescription for experimental determination of the dynamics of a quantum black box. *Journal of Modern Optics*, **44**, 2455.

[46] Cirac, J. I., Ekert, A. K., Huelga, S. F., et al., 1999. Distributed quantum computation over noisy channels. *Physical Review A*, **59**, 1999.

[47] Cohen-Tannoudji, C., Dupont-Roc, J. and Grynberg, G. 1998. *Atom–Photon Interactions: Basic Processes and Applications*. 1st ed. Wiley-Interscience, Germany.

[48] Cormen, T. H., Leiserson, C. E., Rivest, R. L., et al., 2009. *Introduction to Algorithms*. MIT Press, Cambridge, MA.

[49] Dean, J. and Ghemawat, S. 2008. MapReduce: simplified data processing on large clusters. *Communications of the ACM*, **51**, 107.

[50] Deutsch, D. 1985. Quantum theory, the Church-Turing principle and the universal quantum computer. *Proceedings of the Royal Society of London A*, **400**, 97.

[51] Deutsch, D., Ekert, A., Jozsa, R., et al., 1996. Quantum privacy amplification and the security of quantum cryptography over noisy channels. *Physical Review Letters*, **77**, 2818.

[52] Deutsch, D. and Jozsa, R. 1992. Rapid solution of problems by quantum computation. *Proceedings of the Royal Society of London A*, **439**, 553.

[53] Devitt, S. J., Munro, W. J. and Nemoto, K. 2013. Quantum error correction for beginners. *Reports on Progress in Physics*, **76**, 076001.

[54] Didier, N., Pugnetti, S., Blanter, Y. M., et al., 2014. Quantum transducer in circuit optomechanics. *Solid State Communications*, **198**, 61.

[55] Dijkstra, E. W. 1959. A note on two problems in connection with graphs. *Numerische Mathematik*, **1**, 269.

[56] Dowling, J. P. 2008. Quantum optical metrology – the lowdown on high-NOON states. *Contemporary Physics*, **49**, 125.

[57] Duan, L.-M., Giedke, G., Cirac, J. I., et al., 2000. Entanglement purification of Gaussian continuous variable quantum states. *Physical Review Letters*, **84**, 4002.

[58] Duan, L.-M., Lukin, M. D., Cirac, J. I., et al., 2001. Long-distance quantum communication with atomic ensembles and linear optics. *Nature*, **414**, 413.

[59] Duan, L.-M., Lukin, M. D., Cirac, J. I., et al., 2001. Long-distance quantum communication with atomic ensembles and linear optics. *Nature*, **414**, 413.

[60] Dunjko, V., Kashefi, E. and Leverrier, A. 2012. Blind quantum computing with weak coherent pulses. *Physical Review Letters*, **108**, 200502.

[61] Dür, W. and Briegel, H. J. 2007. Entanglement purification and quantum error correction. *Reports on Progress in Physics*, **70**, 1381.

[62] Dür, W., Briegel, H. J., Cirac, J. I., et al., 1999. Quantum repeaters based on entanglement purification. *Physical Review A*, **59**, 169.

[63] Einstein, A., Podolsky, B. and Rosen, N. 1935. Can quantum-mechanical description of physical reality be considered complete? *Physical Review*, **47**, 777.

[64] Enk, S., Cirac, J. I. and Zoller, P. 1998. Photonic channels for quantum communication. *Science*, **279**, 205.

[65] Feynman, R. P. 1985. Quantum mechanical computers. *Foundations of Physics*, **16**, 507.

[66] Fishburn, P. C. 1970. *Utility Theory for Decision Making*. John Wiley & Sons, New York.

[67] Fitch, M. J., Jacobs, B. C., Pittman, T. B., et al., 2003. Photon number resolution using time-multiplexed single-photon detectors. *Physical Review A*, **68**, 043814.

[68] Fowler, A. G., Wang, D. S., Hill, C. D., et al., 2010. Surface code quantum communication. *Physical Review Letters*, **104**, 180503.

[69] Fowler, A. G., Mariantoni, M., Martinis, J. M., et al., 2012. Surface codes: Towards practical large-scale quantum computation. *Physical Review A*, **86**, 032324.

[70] Fredman, M. L. and Tarjan, R. E. 1984. Fibonacci heaps and their uses in improved network optimization algorithms. *Proceedings of the 25th IEEE Symposium on the Foundations of Computer Science*, **346**, 338.

[71] Gács, P. 1983. Reliable computation with cellular automata. *Proceedings of the ACM Symposium on Theory of Computing*, **15**, 32.

[72] Gentry, C. 2009. Fully homomorphic encryption using ideal lattices. *Proceedings of the 41st Annual ACM Symposium on Theory of Computing*, **41**, 169.

[73] Gerry, C. C. and Knight, P. L. 2005. *Introductory Quantum Optics*. Cambridge University Press, London.

[74] Gilchrist, A., K., Langford, N. and Nielsen, M. A. 2005. Distance measures to compare real and ideal quantum processes. *Physical Review A*, **71**, 062310.

[75] Gimeno-Segovia, M., Shadbolt, P., Browne, D. E., et al., 2015. From three-photon Greenberger-Horne-Zeilinger states to ballistic universal quantum computation. *Physical Review Letters*, **115**(2), 020502.

[76] Gisin, N., Ribordy, G., Tittel, W., et al., 2002. Quantum cryptography. *Reviews in Modern Physics*, **74**, 145.

[77] Gisin, N. and Thew, R. 2007. Quantum communication. *Nature Photonics*, **1**, 165.

[78] Goebel, A. M., Wagenknecht, G., Zhang, Q., et al., 2008. Multistage entanglement swapping. *Physical Review Letters*, **101**, 080403.

[79] Gottesman, D. 1997. Stabilizer Codes and Quantum Error Correction (PhD thesis, Caltech). *quant-ph/9705052*.

[80] Gottesman, D. and Chuang, I. L. 1999. Demonstrating the viability of universal quantum computation using teleportation and single-qubit operations. *Nature*, **402**, 390.

[81] Greenberger, D. M., Horne, M. A. and Zeilinger, A. 1989. Going beyond Bell's theorem. Page 69–72 of: *Bell's Theorem, Quantum Theory, and Conceptions of the Universe*, M. Kafatos, ed. Kluwer Academic, Dordrecht, The Netherlands.

[82] Gross, D., Kieling, K. and Eisert, J. 2006. Potential and limits to cluster state quantum computing using probabilistic gates. *Physical Review A*, **74**, 042343.

[83] Grover, L. K. 1996. A fast quantum mechanical algorithm for database search. Page 212–219 of: *Proceedings of the 28th Annual ACM Symposium on Theory of Computing*. ACM, Philadelphia. https://doi.org/10.1145/237814.237866

[84] Harrow, A. W., Hassidim, A. and Lloyd, S. 2009. Quantum algorithm for linear systems of equations. *Physical Review Letters*, **103**, 150502.

[85] Hart, P. E., Nilsson, N. J. and Raphael, B. 1968. A formal basis for the heuristic determination of minimum cost paths. *IEEE Transactions on Systems, Man, and Cybernetics.*, **4**, 100.

[86] Hill, C. D., Peretz, E., Hile, S. J., co-authors, 2015. A surface code quantum computer in silicon. *Science Advances*, **1**(9).

[87] Holevo, A. S. 1998. The capacity of the quantum channel with general signal states. *IEEE Transactions on Information Theory*, **44**, 269.

[88] Hong, C. K., Ou, Z. Y. and Mandel, L. 1987. Measurement of sub-picosecond time intervals between two photons by interference. *Physical Review Letters*, **59**, 2044.

[89] Hwang, W.-Y. 2003. Quantum key distribution with high loss: Toward global secure communication. *Physical Review Letters*, **91**, 057901.

[90] Imamoğlu, A. 2009. Cavity QED based on collective magnetic dipole coupling: Spin ensembles as hybrid two-level systems. *Physical Review letters*, **102**, 083602.

[91] Jain, N., Stiller, B., Khan, I., et al., 2016. Attacks on practical quantum key distribution systems (and how to prevent them). *Contemporary Physics*, **57**, 366.

[92] Jiang, L., Taylor, J. M., Nemoto, K., et al., 2009. Quantum repeater with encoding. *Physical Review A*, **79**, 032325.

[93] Jones, N. C., van Meter, R., Fowler, A. G., co-authors, 2012. Layered architecture for quantum computing. *Physical Review X*, **2**(3), 031007.

[94] Jordan, S. P., Lee, K. S. M. and Preskill, J. 2012. Quantum algorithms for quantum field theories. *Science*, **336**, 1130.

[95] Kaltenbaek, R., Aspelmeyer, M., Jennewein, T., co-authors, 2004. Proof-of-concept experiments for quantum physics in space. Page 17 of: *Proceedings of the SPIE, Quantum Communications and Quantum Imaging*, Vol. 5161. SPIE. https://doi.org/10.1117/12.506979

[96] Kieling, K., Gross, D. and Eisert, J. 2007. Minimal resources for linear optical one-way computing. *Journal of the Optical Society of America B*, **24**(2), 184–188.

[97] Kieling, K., Gross, D. and Eisert, J. 2007. Cluster state preparation using gates operating at arbitrary success probabilities. *New Journal of Physics*, **9**, 200.

[98] Kieling, K., Rudolph, T. and Eisert, J. 2006. Percolation, renormalization, and quantum computing with non-deterministic gates. *Physical Review Letters*, **99**, 130501.

[99] Kimble, H. J. 2008. The quantum internet. *Nature*, **453**, 1023.

[100] Kiraz, A., Atatüre, M. and Imamoğlu, A. 2004. Quantum-dot single-photon sources: Prospects for applications in linear optics quantum-information processing. *Physical Review A*, **69**, 032305.

[101] Kitaev, A. Y. 1997. Quantum computations: Algorithms and error correction. *Russian Mathematical Surveys*, **52**(6), 1191.

[102] Knill, E. 2005. Quantum computing with realistically noisy devices. *Nature*, **434**, 39.

[103] Knill, E. and Laflamme, R. 1997. Theory of quantum error-correcting codes. *Physical Review A*, **55**, 900.

[104] Kurtsiefer, C., Zarda, P., Mayer, S., et al., 2001. The breakdown flash of silicon avalanche photodiodes—Back door for eavesdropper attacks? *Journal of Modern Optics*, **48**, 2039.

[105] Kwiat, P. G., Mattle, K., Weinfurter, H., et al., 1995. New high-intensity source of polarization-entangled photon pairs. *Physical Review Letters*, **75**, 4337.

[106] Lekitsch, B., Weidt, S., Fowler, A. G., co-authors, 2017. Blueprint for a microwave trapped ion quantum computer. *Science Advances*, **3**(2).

[107] Lloyd, S. 1996. Universal quantum simulators. *Science*, **273**, 1073.

[108] Lloyd, S., Garnerone, S. and Zanardi, P. 2016. Quantum algorithms for topological and geometric analysis of data. *Nature Communications*, **7**, 10138.

[109] Lloyd, S., Mohseni, M. and Rebentrost, P. 2013. Quantum algorithms for supervised and unsupervised machine learning. Preprint, arXiv:1307.0411.

[110] Lo, H.-K., Ma, X. and Chen, K. 2005. Decoy state quantum key distribution. *Physical Review Letters*, **94**, 230504.

[111] Menicucci, N. C., Baragiola, B. Q., Demarie, T. F., et al., 2018. Anonymous broadcasting of classical information with a continuous-variable topological quantum code. *Physical Review A*, **97**, 032345.

[112] Metodiev, T., Cross, A., Thaker, D., co-authors, 2004. Preliminary results on simulating a scalable fault-tolerant ion trap system for quantum computation. In: *3rd Workshop on Non-Silicon Computing (NSC-3)*, Munich.

[113] Morimae, T., Dunjko, V. and Kashefi, E. 2015. Ground state blind quantum computation on AKLT state. *Quantum Information and Computation*, **15**, 0200.

[114] Morimae, T. and Fujii, K. 2013. Blind topological measurement-based quantum computation. *Physical Review A*, **87**, 050301(R).

[115] Motes, K. R., Dowling, J. P., Gilchrist, A., et al., 2015. Implementing scalable Boson sampling with time-bin encoding: Analysis of loss, mode mismatch, and time jitter. *Physical Review A*, **92**, 052319.

[116] Motes, K. R., Olson, J. P., Rabeaux, E. J., et al., 2015. Linear optical quantum metrology with single photons: Exploiting spontaneously generated entanglement to beat the shot-noise limit. *Physical Review Letters*, **114**, 170802.

[117] Mukai, H., Sakata, K., Devitt, S. J., co-authors, 2020. Pseudo-2D superconducting quantum computing circuit for the surface code: Proposal and preliminary tests. *New Journal of Physics*, **22**(4), 043013.

[118] Munro, W. J., Azuma, K., Tamaki, K., et al., 2015. Inside quantum repeaters. *IEEE Journal of Selected Topics in Quantum Electronics*, **21**, 6400813.

[119] Munro, W. J., Harrison, K. A., Stephens, A. M., et al., 2010. From quantum multiplexing to high-performance quantum networking. *Nature Photonics*, **4**, 792.

[120] Munro, W. J., Van Meter, R. Louis, S. G. R., et al., 2008. High-bandwidth hybrid quantum repeater. *Physical Review Letters*, **101**, 040502.

[121] Munro, W. J., Stephens, A. M., Devitt, S. J., et al., 2012. Quantum communication without the necessity of quantum memories. *Nature Photonics*, **6**, 777.

[122] Muralidharan, S., Kim, J., Lütkenhaus, N., et al., 2014. Ultrafast and fault-tolerant quantum communication across long distances. *Physical Review Letters*, **112**, 250501.

[123] Muralidharan, S., Li, L., Kim, J., 2015. Optimal architectures for long distance quantum communication. *Scientific Reports*, **6**, 20463.

[124] Nemoto, K., Trupke, M., Devitt, S. J., co-authors, 2014. Photonic architecture for scalable quantum information processing in diamond. *Physical Review X*, **4**(3), 031022.

[125] Nielsen, M. A. 2004. Optical quantum computation using cluster states. *Physical Review Letters*, **93**, 040503.

[126] Nielsen, M. A. 2006. Cluster-state quantum computation. *Reviews in Mathematical Physics*, **57**, 147.

[127] Nielsen, M. A. and Chuang, I. L. 2000. *Quantum Computation and Quantum Information*. Cambridge University Press, Cambridge, UK.

[128] O'Brien, J. L., Pryde, G. J., Gilchrist, A., et al., 2004. Quantum process tomography of a controlled-NOT gate. *Physical Review Letters*, **93**, 080502.

[129] Oxborrow, M. and Sinclair, A. G. 2005. Single-photon sources. *Contemporary Physics*, **46**, 173.

[130] Pan, J.-W., Gasparoni, S., Ursin, R., et al., 2003. Experimental entanglement purification of arbitrary unknown states. *Nature*, **423**, 417.

[131] Pan, J.-W., Simon, C., Brukner, Č., et al., 2001. Entanglement purification for quantum communication. *Nature*, **410**, 1067.

[132] Pirandola, S., Andersen, U. L., Banchi, L., co-authors, 2019. Advances in Quantum Cryptography. *arXiv preprint arXiv:1906.01645*.

[133] Poundstone, W. 1993. *Prisoner's Dilemma/John von Neumann, Game Theory and the Puzzle of the Bomb*. Anchor, Palatine, IL.

[134] Rabl, P., Kolkowitz, S. J., Koppens, F. H. L., et al., 2010. A quantum spin transducer based on nanoelectromechanical resonator arrays. *Nature Physics*, **6**, 602.

[135] Raimond, J.-M., Brune, M. and Haroche, S. 2001. Manipulating quantum entanglement with atoms and photons in a cavity. *Reviews in Modern Physics*, **73**, 565.

[136] Ralph, T. C., Hayes, A. and Gilchrist, A. 2005. Loss-tolerant optical qubits. *Physical Review Letters*, **95**, 100501.

[137] Raussendorf, R. and Briegel, H. J. 2001. A one-way quantum computer. *Physical Review Letters*, **86**, 5188.

[138] Raussendorf, R., Browne, D. E. and Briegel, H. J. 2003. Measurement-based quantum computation on cluster states. *Physical Review A*, **68**, 022312.

[139] Reck, M., Zeilinger, A., Bernstein, H. J., et al., 1994. Experimental realization of any discrete unitary operator. *Physical Review Letters*, **73**, 58.

[140] Rivest, R. L., Adleman, L. and Dertouzos, M. L. 1978. On data banks and privacy homomorphisms. *Foundations of Secure Computation*, 4(11), 169–180.

[141] Rivest, R. L., Shamir, A. and Adleman, L. 1978. A method for obtaining digital signatures and public-key cryptosystems. *Communications of the ACM*, **21**, 120.

[142] Rohde, P. P. 2012. Optical quantum computing with photons of arbitrarily low fidelity and purity. *Physical Review A*, **86**, 052321.

[143] Rohde, P. P. 2015. Boson-sampling with photons of arbitrary spectral structure. *Physical Review A*, **91**, 012307.

[144] Rohde, P. P. 2015. A simple scheme for universal linear optics quantum computing with constant experimental complexity using fiber-loops. *Physical Review A*, **91**, 012306.

[145] Rohde, P. P. and Barrett, S. D. 2007. Strategies for the preparation of large cluster states using non-deterministic gates. *New Journal of Physics*, **9**, 198.

[146] Rohde, P. P., Fitzsimons, J. F. and Gilchrist, A. 2013. The information capacity of a single photon. *Physical Review A*, **88**, 022310.

[147] Rohde, P. P., Helt, L. G., Steel, M. J., et al., 2015. Multiplexed single-photon state preparation using a fibre-loop architecture. *Physical Review A*, **92**, 053829.

[148] Rohde, P. P., Mauerer, W. and Silberhorn, C. 2007. Spectral structure and decompositions of optical states, and their applications. *New Journal of Physics*, **9**, 91.

[149] Rohde, P. P., Pryde, G. J., O'Brien, J. L., et al., 2005. Quantum-gate characterization in an extended Hilbert space. *Physical Review A*, **72**, 032306.

[150] Rohde, P. P. and Ralph, T. C. 2005. Frequency and temporal effects in linear optical quantum computing. *Physical Review A*, **71**, 032320.

[151] Rohde, P. P. and Ralph, T. C. 2006. Error models for mode-mismatch in linear optics quantum computing. *Physical Review A*, **73**, 062312.

[152] Rohde, P. P. and Ralph, T. C. 2011. Time-resolved detection and mode-mismatch in a linear optics quantum gate. *New Journal of Physics*, **13**, 053036.

[153] Rohde, P. P., Ralph, T. C. and Munro, W. J. 2006. Practical limitations in optical entanglement purification. *Physical Review A*, **73**, 030301(R).

[154] Rohde, P. P., Ralph, T. C. and Munro, W. J. 2007. Error tolerance and tradeoffs in loss- and failure-tolerant quantum computing schemes. *Physical Review A*, **75**, 010302(R).

[155] Rohde, P. P., Ralph, T. C. and Nielsen, M. A. 2005. Optimal photons for quantum information processing. *Physical Review A*, **72**, 052332.

[156] Rohde, P. P., Webb, J. G., Huntington, E. H., et al., 2007. Comparison of architectures for approximating number-resolving photo-detection using non-number-resolving detectors. *New Journal of Physics*, **9**, 233.

[157] Sakurai, J. J. 1994. *Modern Quantum Mechanics*. Addison-Wesley, Reading, MA.

[158] Sangouard, N., Simon, C., de Riedmatten, H., et al., Quantum repeaters based on atomic ensembles and linear optics. *Reviews in Modern Physics*, **83**, 33.

[159] Santori, C., Pelton, M., Solomon, G., et al., 2001. Triggered single photons from a quantum dot. *Physical Review Letters*, **86**, 1502.

[160] Scarani, V., Bechmann-Pasquinucci, H. Cerf, N. J., et al., 2009. The security of practical quantum key distribution. *Reviews in Modern Physics*, **81**, 1301.

[161] Scheidl, T., Wille, E. and Ursin, R. 2013. Quantum optics experiments using the International Space Station: A proposal. *New Journal of Physics*, **15**, 043008.

[162] Schneier, B. 1996. *Applied Cryptography*. John Wiley & Sons, Hoboken, NJ.

[163] Schuetz, M. J. A., Kessler, E. M., Giedke, G., et al., 2015. Universal quantum transducers based on surface acoustic waves. *Physical Review X*, **5**, 031031.

[164] Schumacher, B. and Westmoreland, M. D. 1997. Sending classical information via noisy quantum channels. *Physical Review A*, **56**, 131.

[165] Shor, P. W. 1994. Algorithms for quantum computation: discrete logarithms and factoring. Page 124 of: *Symposium on the Foundations of Computer Science*, Vol. 35. IEEE.

[166] Shor, P. W. 1995. Scheme for reducing decoherence in quantum computer memory. *Physical Review A*, **52**, R2493.

[167] Shumeiko, V. S. 2016. Quantum acousto-optic transducer for superconducting qubits. *Physical Review A*, **93**, 023838.

[168] Stannigel, K., Rabl, P., Sørensen, A. S., et al., 2010. Optomechanical transducers for long-distance quantum communication. *Physical Review Letters*, **105**, 220501.

[169] Stephens, A. M. 2014. Fault-tolerant thresholds for quantum error correction with the surface code. *Physical Review A*, **89**, 022321.

[170] Stephens, A. M., Fowler, A. G. and Hollenberg, L. C. L. 2008. Universal fault-tolerant computation on bilinear nearest neighbor arrays. *Quantum Information and Computation*, **8**, 330.

[171] Stephens, A. M., Huang, J., Nemoto, K., et al., 2013. Hybrid-system approach to fault-tolerant quantum communication. *Physical Review A*, **87**, 052333.

[172] Straffin, P. D. 1993. *Game Theory and Strategy*. Mathematical Association of America, Washington, DC.

[173] Sugden, R. 2004. *The Economics of Rights, Co-operation and Welfare*. Palgrave Macmillan, New York.

[174] Svore, K. M., DiVincenzo, D. P. and Terhal, B. M. 2007. Noise threshold for a fault-tolerant two-dimensional lattice architecture. *Quantum Informatics and Computation*, **7**, 297.

[175] Szkopek, T., Boykin, P. O., Fan, H., co-authors, 2006. Threshold error penalty for fault-tolerant computation with nearest neighbour communication. *IEEE Transactions on Nanotechnology*, **5**(1), 42.

[176] Tan, S.-H. and Rohde, P. P. 2019. The resurgence of the linear optics quantum interferometer – Recent advances & applications. *Reviews in Physics*, **4**, 100030.

[177] Tanenbaum, A. S. 2002. *Computer Networks*. Prentice Hall, Hoboken, NJ.

[178] U'Ren, A. B., Banaszek, K. and Walmsley, I. A. 2003. Photon engineering for quantum information processing. *Quantum Information & Computation*, **3**, 480.

[179] U'Ren, A. B., Silberhorn, C., Banaszek, K., co-authors, 2005. Generation of pure-state single-photon wavepackets by conditional preparation based on spontaneous parametric downconversion. *Laser Physics*, **15**, 146.

[180] Van Dijk, M., Gentry, C., Halevi, S., et al., 2010. Fully homomorphic encryption over the integers. *Advances in Cryptology – EUROCRYPT*, 24. Springer, Berlin.

[181] Van Loock, P., Ladd, T. D., Sanaka, K., et al., 2006. Hybrid quantum repeater using bright coherent light. *Physical Review Letters*, **96**, 240501.

[182] Van Meter, R. 2014. *Quantum Networking*. Wiley, Hoboken, NJ.

[183] Vinay, S. E. and Kok, P. 2018. Extended analysis of the Trojan-horse attack in quantum key distribution. *Physical Review A*, **97**, 042335.

[184] von Neumann, J. 1955. Probabilistic logics and the synthesis of reliable organisms from unreliable components. *Automata Studies*, **43**.

[185] von Neumann, J. and Morgenstern, O. 2007. *Theory of Games and Economic Behavior*. Princeton University Press, Princeton, NJ.

[186] Wallraff, A., Schuster, D. I., Blais, A., co-authors, 2004. Strong coupling of a single photon to a superconducting qubit using circuit quantum electrodynamics. *Nature*, **431**, 162.

[187] Wang, D. S., Fowler, A. G. and Hollenberg, L. C. L. 2011. Quantum computing with nearest neighbor interactions and error rates over 1%. *Physical Review A.*, **83**, 020302(R).

[188] Wang, D. S., Fowler, A. G., Stephens, A. M., et al., 2010. Threshold error rates for the toric and surface codes. *Quantum Informatics and Computation*, **10**, 456.

[189] Yin, J., Cao, Y., Li, Y.-H., co-authors, 2017. Satellite-based entanglement distribution over 1200 kilometers. *Science*, **356**, 1140.

[190] Yin, J., Cao, Y., Yong, H.-L., et al., 2013. Lower bound on the speed of nonlocal correlations without locality and measurement choice loopholes. *Physical Review Letters*, **110**, 260407.

[191] Yoran, N. and Reznik, B. 2003. Deterministic linear optics quantum computation with single photon qubits. *Physical Review Letters*, **91**, 037903.

[192] Zehnder, L. 1891. Ein neuer Interferenzrefraktor [A new interference refractor]. *Zeitschrift für Instrumentenkunde*, **11**, 275.

[193] Zehnder, L. 1892. Über einen Interferenzrefraktor [On an interference refractor]. *Zeitschrift für Instrumentenkunde*, **12**, 89.

[194] Zukowski, M., Zeilinger, A., Horne, M. A., et al., 1993. Event-ready-detectors Bell experiment via entanglement swapping. *Physical Review Letters*, **71**, 4287.

Index